Living offshore reefs of
Australian Marine Parks

Reef Life Survey acknowledge the Traditional Owners of Country throughout Australia, their continuing connection to land, sea, sky and community, and their ongoing cultural rights and responsibilities. We pay our respects to them and their cultures and to their elders past and present. We recognise that Traditional Owners have been using and managing their Sea Country, including areas now included within Australian Marine Parks, for thousands of years, in some cases since before rising sea levels created these marine environments.

Living offshore reefs of
Australian Marine Parks

Graham Edgar, Rick Stuart-Smith and Antonia Cooper

Reef Life Survey Foundation

REEF LIFE
SURVEY

Contents

Introduction

Australia's living marine heritage

As an island continent, Australia is uniquely privileged in possessing extensive tropical and temperate reef ecosystems. Australia also has a uniquely important global responsibility to safeguard marine biodiversity, given its position amongst wealthy developed countries, relatively low human population pressure, and healthy populations of many species that are threatened elsewhere.

All Australians recognise and appreciate the environmental values of coral reefs, as represented by the Great Barrier Reef – incredibly rich communities of plants and animals that support fishing, tourism, recreation, cultural practices, education, and general sense of well-being. Less appreciated, but no less important, is the 'Great Southern Reef', the ecologically connected network of rocky reefs that extends around Australia's southern coastlines[1]. Southern reefs are home to tens of thousands of temperate species that live only on this continent, and thus cannot be protected by other nations. Also notable is that many temperate species possess long evolutionary lineages, a consequence of geographic isolation for 55 million years since fragmentation and separation from Gondwana. An example is the handfish family, with 14 relict species now persisting in Tasmania and south-eastern Australia only, but once with a global distribution, as revealed by 50-million-year-old European fossils[2].

Australian reefs have been impacted by numerous human pressures. High population density on coastal plains has led to expansion of numerous industries and an associated cocktail of stressors that affect reef ecosystems[3-5]. Nevertheless, human pressures are distributed patchily, and many reefs are situated far from towns and cities.

3 images above: Dead coral habitat after 2016 heatwave. Osprey Reef, Coral Sea Marine Park.

The more important threats to Australia's reefs are:[6]

1. Climate change. This overarching threat has already dramatically changed some reef habitats, including through the transformation of kelp forests to barren grounds, and by bleaching and subsequent loss of coral. Species respond to changing climate – rising sea temperatures, increasing seawater acidity, changing sea level, altered rainfall patterns, and increased cyclone risk – by moving or by population abundance change. These lead to the establishment of 'novel communities', groups of interacting species that previously did not mix.

2. Runoff from land clearance in catchments. Massive quantities of silt transported down creeks, rivers and stormwater drains can smother inshore reefs, particularly after floods. Nutrients draining off farmland and transported down rivers cause algal blooms, which reduce dissolved oxygen and allow fine algae to overgrow and smother kelps, seagrasses, corals, sponges and other attached animals[7].

3. Fishing. Populations of fished species decline under fishing pressure. In addition, perhaps even more importantly, fishing affects species caught incidentally as 'by-catch', species that lose critical habitat structure through trawling, and species affected by changes to food webs where sharks, groupers and other top predators are removed[8].

4. Introduced species. New arrivals threaten native species and functioning of ecosystems through habitat modification, competition, predation, disease, and poisoning[9,10]. This threat is global, largely uncontrollable, and accelerating.

5. Chemical and nutrient pollution. Oil spills, fish farm waste, land reclamation, foreshore development, sewage effluent, heavy metal discharge, and chemical outfalls generate intense localised impacts on the marine environment.

6. Shipping and military operations. Shipping affects reefs in many ways, including through oil spills, iron pollution (see p. 145), release of toxic anti-fouling chemicals, introduced species, noise, and direct

Bleached *Pocillopora* coral. Osprey Reef, Coral Sea Marine Park.

abrasion of the seabed. Military activities are often excluded from environmental controls, adding to the sum of impacts across the seascape.

7. Plastics and other debris. Discarded netting ('ghost nets') can continue to catch and kill turtles, seabirds, marine mammals and fishes for years after loss. Floating plastics are consumed by seabirds, turtles and fishes, causing death in some instances[11]. Microscopic plastic particles are ingested in large quantities by many sea creatures, accumulating in the gut and body tissues, and clogging gills.

8. Noise pollution. Human activities contribute massively to undersea noise. This interferes with communication, balance, behaviour, and homing of sea creatures[12].

The various environmental changes in the marine environment are occurring out-of-sight below the sea surface. Consequently the magnitude and distribution of change remains largely unknown. Unlike an oil spill or bushfire that attract immediate attention, many serious threats to marine species progress slowly, eroding species' populations over time-scales of decades to centuries[13].

We do know, however, that many marine habitats are deteriorating. This is particularly the case for estuaries and shallow reefs. Estuaries provide the focal areas for human settlement. Shallow reefs are subjected to many overlapping threats that interact with each other in largely unknown ways.

Large steel shipwrecks and isolated reefs were used historically for military target practice. Unexploded ordnance, such as this shell near the wreck of the *Runic* at Middleton Reef in the Lord Howe Marine Park, pose a risk for both divers and marine life.

Reef Life Survey (RLS)

With a shared concern that the marine environment was deteriorating out-of-sight, a group of divers, managers and scientists agreed to collaborate to accurately assess the health of Australia's reefs. We recognised that scientific knowledge was localised, with research effort concentrated in small areas, which may or may not be typical of the wider region. Management efforts were focused on threats attracting public attention, which may not necessarily be the primary drivers of environmental change. Collecting data in locations all around Australia using the same methodology would enable national comparisons to be made and hence the capacity to identify in which locations populations were declining most rapidly, and which were the most critical threats to be addressed.

Thus, in 2007 Reef Life Survey Foundation (RLS; reeflifesurvey.com) was born. We aspired to use the skills and enthusiasm of volunteer divers to conduct rigorous surveys of marine life across scales impossible for scientific teams to cover, and to provide critically needed information for managers to prioritise actions to address threats. Over 300 RLS divers have now been trained to swim up and down transect lines around the country, recording numbers of fishes, turtles, sea snakes, sea stars, lobsters, abalone and other invertebrates, and completing photographic surveys of plants, corals, sponges and other animals attached to the seabed. Management agencies actively participate through the Advisory Committee, which includes representatives from national, state and territory environment departments, and diving groups.

Our five goals are to:

1. Design and apply a standardised methodology for surveying shallow marine life from the tropics to the poles.
2. Identify, train and support a community of citizen scientists comprising the most capable recreational divers in the application of visual survey methods.
3. Collate, curate and freely distribute data online.
4. Develop regular communication pathways between scientists, volunteer divers and managers, all committed to improving marine conservation outcomes.
5. Communicate marine environmental knowledge to the wider public.

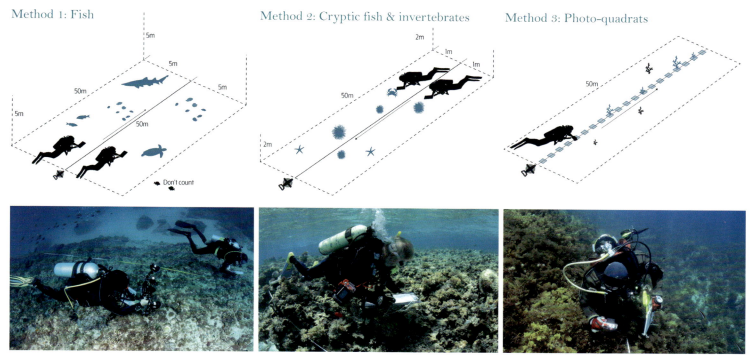

Method 1: Fish   Method 2: Cryptic fish & invertebrates   Method 3: Photo-quadrats

Survey methods used by Reef Life Survey divers.

Methods used on RLS surveys centre on a 50m transect tape swam out by divers, as described in the RLS standardised methods manual[14]. All fishes are firstly counted in 5m-wide blocks each side of the transect line. Mobile invertebrates and cryptic fishes are then counted in 1m-wide blocks. Finally, the density of attached organisms such as coral and kelp is documented by taking photographs of the seabed in 20 positions along the line. The transect is then repeated elsewhere on the reef.

RLS collects data in four ways: 1. Trained divers complete surveys on their own initiative at local reefs and while on holidays. 2. RLS Foundation schedules annual surveys at core monitoring locations distributed around Australia, where divers gather as a group to resurvey long-term monitoring sites. 3. Sites are surveyed from a sailing catamaran over 2–4 week periods with volunteer crew, allowing access to remote offshore Australian Marine Park sites. 4. University, not-for-profit and management agency staff conduct scientific surveys using the RLS methodology.

Great emphasis is placed on maintaining high data standards. Interested divers firstly need to be keen and experienced (more than 50 dives), then are trained one-on-one by qualified trainers over at least eight survey dives. Volunteers need to exceed data-quality benchmarks to complete training, with ongoing comparative checks of data and images by scientific experts. Training is assisted by the online 'Frequency Explorer' tool (reeflifesurvey.com/frequency-explorer), which provides species ranked by abundance with images for any region, and includes automated flashcards for testing divers' knowledge. Data from underwater sheets are transcribed into digital form on data entry templates that flag common data entry errors. The data are further screened for errors during input to the central database.

TONI COOPER

TONI COOPER

The fundamental attribute of divers participating in Reef Life Survey is their enormous dedication. Why bother? Most stay engaged primarily because they see RLS efforts contributing in a concrete way to improving the marine environment. An internal survey of RLS divers revealed that most participate to 'contribute to marine science and management' (89 per cent of RLS divers), to 'further knowledge of marine life' (83 per cent), and to 'contribute to conservation in general' (66 per cent).

So far, RLS volunteers have counted numbers of more than 4,000 animal species along 15,000 transects at over 3,500 sites – not just around Australia but also in 52 other countries from the Arctic to Antarctica. RLS data now provide the most detailed standardised reference describing population numbers of marine species as they are today – an irreplaceable baseline to be used through the future when assessing how marine ecosystems are changing. RLS data, maps and identification tools can be accessed online, allowing anyone to determine exactly what biodiversity changes are happening underwater, and thus where threats

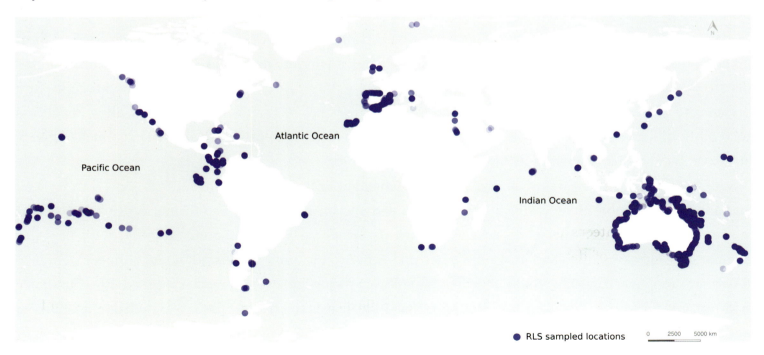

Atlantic Ocean

Pacific Ocean

Indian Ocean

● RLS sampled locations

0   2500   5000 km

Global distribution of sites surveyed by RLS divers.

TONI COOPER

to marine life are concentrated. Through images and data generated by the RLS community and showcased on our 'Reef Species of the World' and 'Reef Explorer' online tools (accessed via reeflifesurvey.com), new information is continually updated on species' locations, population trends and rarity.

In this book, we summarise RLS data for key indicators on environmental condition of shallow reefs within Australian Marine Parks. For each park, the size of the icon reflects the magnitude of the indicator. As depicted for the Cod Grounds Marine Park example below, comparative data are provided on eight indicators: 1. Estimated total biomass (weight) of all fish sighted in a 500m² area surveyed along each transect, then averaged across all transects within the marine park. 2. Average number of fish species sighted per 500m² transect area. 3. Average number of cryptic fish species sighted in the narrower 100m² transect blocks. 4. Average number of mobile invertebrate species recorded per 100m² transect area. 5. Average number of sea snakes sighted per 500m² transect area. 6. Average number of sharks sighted per 500m² transect area. 7. Average number of Crown-of-thorns sea stars (COTs) sighted per 100m² transect area. 8. Average per cent of the seabed covered by corals on transects. To the right of the values given at the bottom of each box is an arrow pointing upward if values in that marine park from earliest to most recent surveys are significantly rising, downward if falling, or are stable (flat line, indicating non-significant change).

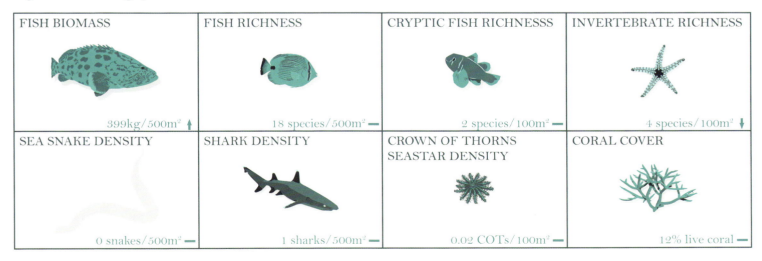

FISH BIOMASS	FISH RICHNESS	CRYPTIC FISH RICHNESSS	INVERTEBRATE RICHNESS
399kg/500m² ↑	18 species/500m² —	2 species/100m² —	4 species/100m² ↓
SEA SNAKE DENSITY	SHARK DENSITY	CROWN OF THORNS SEASTAR DENSITY	CORAL COVER
0 snakes/500m² —	1 sharks/500m² —	0.02 COTs/100m² —	12% live coral —

Australian Marine Parks

Marine protected areas are recognised as one of the best ways to conserve and protect marine habitats and species in our oceans. In light of this, governments Australia-wide agreed in 1998 to establish a National Representative System of Marine Protected Areas (NRSMPA). The NRSMPA aimed to create a comprehensive, adequate and representative system of marine protected areas, to contribute to the long-term viability of the marine environment, and to protect biodiversity.

Steps to establish Australian Marine Parks began in 1982. Today, there are 58 Australian Marine Parks around the country. Management plans for these parks commenced in 2013 and 2018. Parks are located in Commonwealth waters, beginning 3 nautical miles offshore and extending to the edge of the Exclusive Economic Zone, 200 nautical miles offshore. Commonwealth waters also surround Australia's external territories, such as Christmas Island and Cocos-Keeling, Norfolk Island, and Heard and McDonald Islands. Commonwealth waters cover a vast area, from tropical waters bordering the global centre of marine biodiversity in Indonesia, Papua New Guinea and the Solomon Islands, to cold temperate environments south of Tasmania, and waters off subantarctic Macquarie, Heard and McDonald Islands. In total, Australian Marine Parks cover around 2.8 million square kilometres of ocean (parksaustralia.gov.au/marine).

Australian waters are subdivided into major regions (the South-east, Temperate East, Coral Sea, North, North-west and South-west), each hosting areas or features of conservation significance that are afforded different degrees of protection from human extraction using zoning. Zone types vary from excluding all extractive activities, excluding activities that impact and damage the seafloor, to multiple use zones that allow extractive use. The distribution of zones included in the management plans was seen by the government as a balance between the imperatives of conservation and sustainable use. However, conservation and science organisations were critical of the scarcity of National Park (no-fishing) Zones in shallow water habitats where threats are concentrated, and little change to ongoing fishing and mining activities[15,16].

Reefs occur in all six Australian Marine Park regions. The Coral Sea Marine Park and the North and North-west Networks tend to have coral reefs; the Temperate East Network has a combination of coral (for example Elizabeth and Middleton Reefs in Lord Howe Marine Park) and rocky reefs (for example the Cod Grounds Marine Park); and the South-west and South-east Networks have rocky reefs. Only a small percentage of the available seabed in any given network is reef; despite this, they tend to host the bulk of the known species richness in each region. This reflects the role of reefs everywhere in attracting a high abundance of marine life. Despite occupying less than 1 per cent of the ocean's

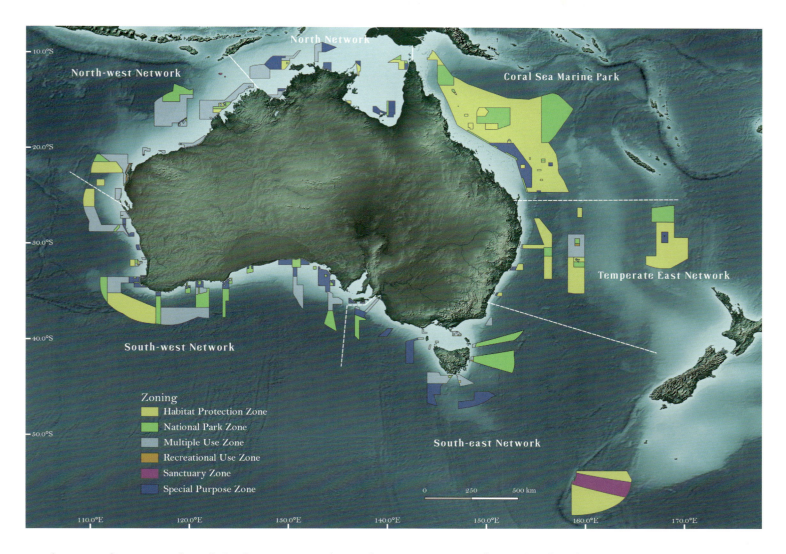

Zoning
- Habitat Protection Zone
- National Park Zone
- Multiple Use Zone
- Recreational Use Zone
- Sanctuary Zone
- Special Purpose Zone

surface, reefs across the globe host approximately 25 per cent of marine biodiversity[17].

Many reefs in Australian Marine Parks lie at shallow depths accessible on SCUBA; this makes reefs the best-studied habitats within Commonwealth waters around Australia. For example, in the Coral Sea Marine Park, coral reefs have attracted about 30 per cent of all the research[18]. Due to their recognised value as critical habitat, many reefs in Australian Marine Parks have a long history of protection. Lihou Reef and the Coringa-Herald group of reefs in the Coral Sea, Ashmore Reef (WA), Mermaid Reef, Middleton Reef, Pimpernel Rock and the Cod Grounds, were all protected from fishing as part of earlier establishment of marine parks in Australia from the 1980s, either because of their near-pristine state or to protect particular resident species[19,20]. The Parks Australia website (parksaustralia.gov.au/marine) provides full details of each Australian Marine Park, including park boundaries, key ecological features, and activities allowed. All photographs in this book were taken by Reef Life Survey divers within Australian Marine Parks, other than the few exceptions where noted from nearby locations.

As this book went to press, new Indian Ocean Territories Marine Parks were declared that include waters surrounding Christmas and Cocos (Keeling) Islands. With these additions, Parks Australia manage 60 marine parks, spanning more than 4 million square kilometres and covering 45 per cent of Australian waters.

North-west Marine Parks Network

The North-west Network extends from the Western Australian/Northern Territory border south-west to near Kalbarri[21]. Conditions are tropical, with a mostly wide continental shelf, a large number of banks and shoals, a highly variable tidal regime, frequent tropical cyclones, and a complex system of ocean currents[22]. The Leeuwin Current and the Indonesian Throughflow greatly affect the region by transporting warm low-nutrient water from the Pacific through the Indonesian archipelago and southward to central WA, and beyond.

The North-West Network covers a total of 335,341km², with habitats that include coral reefs, soft sediments, deep canyons, calcareous pavements, intertidal flats, and sheltered lagoons. The tidal range varies enormously, from spring tides that can exceed 10m in parts of the Kimberley Marine Park in the north to microtides of less than 1m in Shark Bay Marine Park in the south. Marine habitats reflect these tidal extremes; large tides in inshore waters produce turbid conditions with high concentrations of silt deposited on corals and other attached animals, while extensive coral reefs develop in clear oceanic waters.

Aboriginal people continue to assert inherited rights and responsibilities over Sea Country within the North-west Network. Sea Country refers to the areas of the sea that Aboriginal people are particularly affiliated with through their traditional lore and customs. It extends from terrestrial areas into nearshore and offshore waters, and is valued for Indigenous cultural identity, health and wellbeing. Songlines that connect landscape features with immense spiritual significance traverse Sea Country.

Resource use and conservation in the North-west Network are guided by the *North-west Marine Parks Network Management Plan 2018*, which regulates activities within the 13 separate Australian Marine Parks. Protection of waters at Ashmore Reef, Cartier Island and offshore of Ningaloo Reef occurred much earlier, in 1983, 2000 and 1987, respectively. Biodiversity conservation was prioritised in these

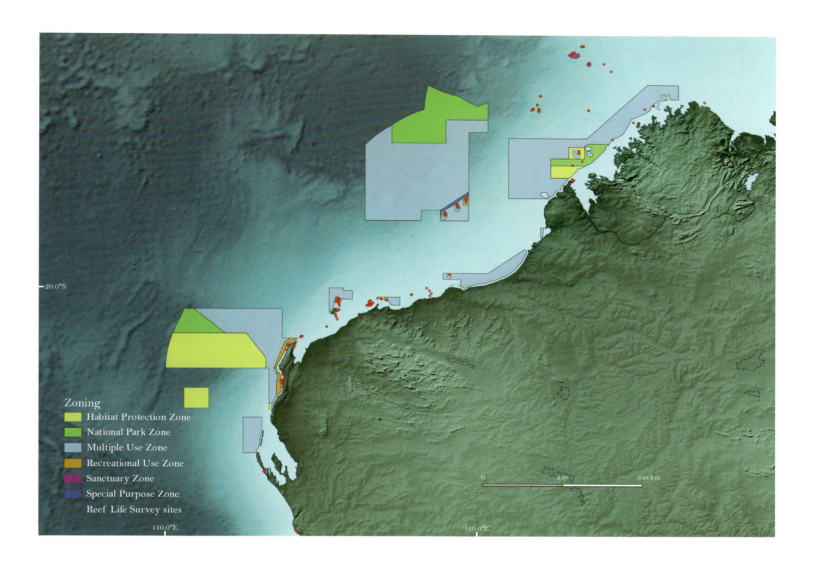

Zoning
- Habitat Protection Zone
- National Park Zone
- Multiple Use Zone
- Recreational Use Zone
- Sanctuary Zone
- Special Purpose Zone
- Reef Life Survey sites

areas due to globally significant marine turtle nesting areas and migratory bird populations, the presence of Dugong, Whale Shark and various cetaceans, and international obligations associated with visiting Indonesian fishers. The islands and surrounding reefs at Ashmore Reef and Cartier Island Marine Parks have been visited for centuries by Indonesian fishers, who collected sea cucumbers, giant clams, trochus and other molluscs, and meat and eggs of birds and turtles, for local consumption and trade on the Asian market.

Reefs in the North-west Network have experienced multiple environmental disturbances in recent decades. Substantial coral bleaching and death occurred during a period of abnormally high sea temperatures in the 1998 summer[23]. However, while some reef systems in the region lost most coral cover, Ashmore and Cartier Island Reefs experienced little initial coral death[23], but subsequent bleaching events may have had a greater impact[24]. Recovery of coral cover appears rapid during years with little disturbance, suggesting high ecological resilience[25].

Ashmore Reef Marine Park

Ashmore Reef is a complex platform reef system 370km north-west of the Kimberley coast but only 210km from Timor, and 120km from Pulau Roti – a heavily populated Indonesian island. The Ashmore Reef Marine Park encompasses a huge range of species and habitats that include sand flats, lagoons, seagrass meadows, shallow reef flats, and deeper offshore reefs, banks and channels. These habitats provide sanctuary for an abundance of sea turtles, reef fishes, sea birds, shore birds, a small population of Dugong, sharks, and other marine species. More coral species have been recorded on reefs around Ashmore than at any other location off the Western Australian coast[26]. This marine park was historically known for its rich and abundant sea snake fauna, however populations have declined. In 2021, several species were re-discovered in the deeper reefs at Ashmore, including the Critically Endangered Short-nosed Sea Snake (see p. 28).

The Ashmore Reef Marine Park is recognised globally as both an Important Bird Area and a significant wetland, with listing in 2003 as a Ramsar Wetland of International Importance. Over 50,000 breeding seabirds congregate each year, competing with each other for space on three tiny islands covering a total area of only 56 hectares, and with limited vegetation. In addition, tens of thousands of shorebirds belonging to 30 species – including Bar-tailed Godwit, Great Knot, and the Critically Endangered Curlew Sandpiper – utilise the shallow sandflats to feed[27]. Foraging grounds within Ashmore Reef Marine Park provide a critical refuelling stop-off during migrations from Asia to Australia.

Ashmore Reef was named after the first European to sight the island, Samuel Ashmore in 1811. Following a dispute with the United States over ownership of phosphate reserves, the island group was annexed by the United Kingdom in 1878, with authority transferred to the Australian Government in 1934. Island resources were heavily exploited historically, firstly by American whalers in the mid-18th century, then by miners removing phosphate deposited by birds as guano (and thus taking away most island topsoil). An Australian Border Force vessel is now stationed at the entrance to the main channel through the reef to provide surveillance and compliance services in northern Australian waters.

Large gorgonian fan (*Melithaea* sp.) filtering food particles from tidal currents in lagoon channel. Ashmore Reef Marine Park.

Above: West Island, the largest of three islands on Ashmore Reef, viewed from the lagoon boat anchorage.

Below left and right: Ashmore Reef has the only shallow lagoon sandflats with seagrass beds within the network of Australian Marine Parks. They are exposed twice a day with the flood and ebb of tides, which can drop over 4m. Predatory fishes follow the tide in across the flats, seeking burrowing invertebrates and baitfish schools.

Shrubby vegetation dominates the small dunes that run behind West Island beaches; dense tufted grasslands occupy the island centre. Seabirds challenge each other for limited nesting space on the highest shrubs.

Juvenile Great Frigatebird (*Fregata minor*). West Island. Ashmore Reef.

Nesting Red-footed Booby (*Sula sula*) amongst shrubs. West Island, Ashmore Reef.

ASHMORE REEF
(-12.25, 123.10)

◎ RLS sites

Zone Name	Zone Area (km²)
Sanctuary Zone (no general access)	550
Recreational Use Zone (limited access for recreational use)	34

FISH BIOMASS	FISH RICHNESS	CRYPTIC FISH RICHNESSS	INVERTEBRATE RICHNESS
50kg/500m² ↑	64 species/500m² —	10 species/100m² —	4 species/100m² —

SEA SNAKE DENSITY	SHARK DENSITY	CROWN OF THORNS SEASTAR DENSITY	CORAL COVER
0 snakes/500m² —	0.15 sharks/500m² —	0 COTs/100m² —	16% live coral —

MoU Box

Traditional Indonesian fisher access to Ashmore Reef Marine Park is guided by a Memorandum of Understanding (MoU) between Australia and Indonesia. The MoU, signed by the Australian and Indonesian Governments in 1974 and reviewed in 1989, sets out arrangements by which traditional fishers may access marine resources in a large sector of the North-west region—the MoU Box. It allows for continued Indonesian traditional fishing activities, thus restricted to sail powered vessels and hand collection. Traditional fishing is limited within Ashmore Reef and Cartier Island. Fishers may only take finfish for immediate consumption in the Recreational Use Zone at Ashmore Reef. They are prohibited from entering Sanctuary Zones unless in the event of an emergency. Elsewhere in the MoU Box (e.g. Scott and Seringapatam reefs), Indonesians using traditional fishing practices may harvest sea cucumber and finfish. Nevertheless, an unintended consequence is that the MoU Box is now open to thousands of Indonesian fishers, including from fishing communities in Sulawesi with no historical interest in the area but attracted since 1980 by reports of plentiful sea cucumbers ('trepang') once their local resources had been depleted[28].

Skeletal woodwork periodically emerges from Ashmore sands – victims of cyclones perhaps, but more likely relics of unlawful fishing, or remnants from a short-lived era when Indonesian vessels transported refugees and illegal migrants the short distance from Pulau Roti. That era closed in September 2001 when the Australian Government excised Ashmore and Cartier Islands from the Australian migration zone. Boats that illegally arrive at Ashmore are impounded and destroyed. Their backbones and ribs provide cover for small fishes on incoming tides.

ANDREW GREEN

Above: Because of high value, large sea cucumbers are now rare outside the Australian Marine Parks in the MoU box that prohibit fishing. Royal Sea Cucumber (*Thelenota anax*), Ashmore Reef.

Below: The Giant Clam (*Tridacna gigas*) is the largest bivalve mollusc, reaching more than 200kg in weight and living for over 100 years. Populations of this shallow-water species were greatly impacted by fishers diving in the Ashmore Reef Marine Park, and recovery following enforcement of marine park regulations has been slow. Overexploitation has caused extinction of the species in many islands within its Indo-Pacific range.

Ashmore Reef Marine Park supports the major regional feeding and breeding grounds for Green Turtles (*Chelonia mydas*), with over 10,000 individuals estimated to feed in the seagrass beds and reef flats year-round. Ashmore Reef Marine Park.

'Being able to dive in one of the most remote locations in Australia, and getting training and experience in surveys of biodiverse coral reefs, is an experience of a lifetime.'
NESTOR ECHEDEY BOSCH GUERRA, RLS BLOG ON VISITING ASHMORE REEF.

Three species of epaulette shark occur on Australian reefs. They are all most easily observed at night while actively searching for prey. When seen crawling over the seabed on their pelvic fins, the origin of their alternative name of 'walking sharks' is immediately clear. Speckled Carpetshark (*Hemiscyllium trispeculare*), Ashmore Reef Marine Park.

Above: Once nightfall descends, the reef transforms. Ashmore Marine Reef Park.

Below: Close proximity to the Coral Triangle adds additional species to reefs that are absent in other Australian waters. The Azure Demoiselle (*Chrysiptera hemicyanea*) has a restricted distribution extending from the North West Shelf to Indonesia. Ashmore Reef Marine Park.

Above: Not just imagination, sunsets are brighter in the Ashmore Reef Marine Park than elsewhere in Australia because of light reflection from dust particles released by Indonesian volcanos. Ashmore Reef Marine Park.

Below: The few fish species specialised for capturing prey at night amongst coral reefs are characterised by unusually large eyes. Below left: Bigeye Soldierfish (*Myripristis pralinia*); Below right: Violet Soldierfish (*Myripristis violacea*), Ashmore Reef Marine Park.

Above left: Rubble and sand seabed within the lagoon support a diverse fauna that tends to be overlooked by divers, including many burrow-dwelling forms. Sixspot Glidergoby (*Valenciennea sexguttata*). Ashmore Reef Marine Park.

Below left: Ten minutes spent watching the Crab-eye Goby (*Signigobius biocellatus*) flitting about coral rubble is ten minutes of joy. Dancing pairs circle each other, signalling continuously by flashing their eyespots in semaphore. Ashmore Reef Marine Park.

Above right: Coral rubble contains abundant food resources for invertebrates such as this Necklace Sea Star (*Fromia monilis*). Ashmore Reef Marine Park.

Below right: Small invertebrate predators roam across rubble bottom, searching for worms and other prey. *Philinopsis gardineri*, Ashmore Reef Marine Park.

The mystery of disappearing sea snakes

Ashmore Reef was famous as one of two global epicentres of sea snake diversity (with the South China Sea); however, populations of all 13 species that once occurred in high numbers catastrophically declined between 2000 and 2004 for reasons that remain debated. Sea snakes are now absent from the sheltered lagoons where once hyper-abundant[29], although underwater video has recently revealed several species, including the presumed locally extinct Short-nosed Sea Snake (*Aipysurus apraefrontalis*), living in the deeper reefs around the island's rim.

Suggestions put forward to explain the loss of species include: (i) an exceptionally strong heatwave or cyclone impacted the snakes, (ii) populations collapsed through disease, (iii) predators increased due to prevention of illegal fishing, and (iv) increased boat traffic with associated pollution negatively affected snake numbers. However, not all hypotheses account for the localised nature of impacts, and the fact that all sea snake species were catastrophically affected. RLS dive surveys revealed large numbers of snakes on all reef systems nearby – Hibernia to the north, Cartier to the south-east, and Scott to the south-west. Impacts of whatever caused sea snake demise were largely localised within the unique and extensive lagoonal shallows at Ashmore. One likely explanation relates to predation by Tiger Sharks (*Galeocerdo cuvier*), a species increasingly sighted in the lagoon. These top predators are attracted to the huge concentration of mating and nesting Green Turtles (*Chelonia mydas*), and are not allowed to be harvested by traditional Indonesian fishers at Ashmore Reef. If a few sharks developed a taste for snakes swimming up to the sea surface to breathe, then the snake population could potentially rapidly decline.

Left: Turtle-headed Sea Snake, (*Emydocephalus annulatus*), Scott Reef.

Opposite above: Dubois' Sea Snake (*Aipysurus duboisii*), Scott Reef.

Opposite below: Olive Sea-Snake (*Aipysurus laevis*), Scott Reef.

Cartier Island Marine Park

Cartier Island was first sighted by a European (Captain Nash) in 1800 while voyaging in his ship *Cartier*. With relatively few birds and thus little guano, this small sandy island was less attractive to colonial powers than Ashmore Reef, so annexed by the British later, in 1909. Sovereignty was transferred to Australia in 1934.

Australia saw this island primarily as a military asset for use as a bombing target until 2000, when marine life was protected through establishment of the Cartier Island Marine Reserve. Access to the island remains prohibited because of the risk of unexploded ordnance. As a long-term Sanctuary Zone (i.e. no-entry), large fish are more prevalent than elsewhere in the North-west including Rowley Shoals; however, shark numbers are much lower than on isolated Pacific reefs in the Coral Sea Marine Park, probably because of high fishing pressure in surrounding waters within the MoU Box. For reasons unknown but most likely associated with recent cyclones and heatwaves, sponges rather than corals dominate space on inshore reefs.

Moray eels are unusually rare on the North West Shelf, including Ashmore Reef and the Oceanic Shoals. More are seen by divers in Cartier Island Marine Park than neighbouring reefs due the diversity of crevices created at the base of soft corals. Juvenile Yellowmargin Moray (*Gymnothorax flavimarginatus*).

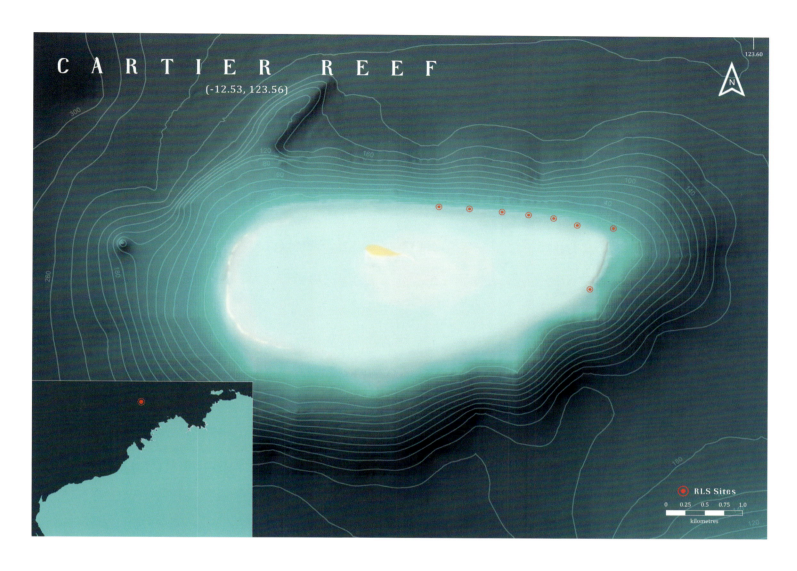

C A R T I E R R E E F

(-12.53, 123.56)

Zone Name	Zone Area (km²)
Sanctuary Zone (no general access)	172

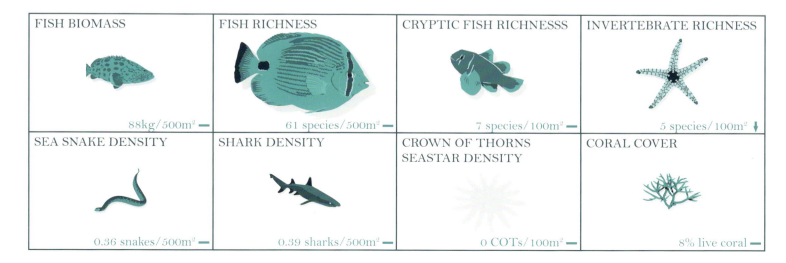

FISH BIOMASS	FISH RICHNESS	CRYPTIC FISH RICHNESSS	INVERTEBRATE RICHNESS
88kg/500m² —	61 species/500m² —	7 species/100m² —	5 species/100m² ↓
SEA SNAKE DENSITY	SHARK DENSITY	CROWN OF THORNS SEASTAR DENSITY	CORAL COVER
0.36 snakes/500m² —	0.39 sharks/500m² —	0 COTs/100m² —	8% live coral —

Above and below: Cartier Island Marine Park.

Above: Diverse sessile invertebrate community dominated by soft corals along the reef edge.
Cartier Island Marine Park.

Above: *Porites* species (left), *Pachyseris speciosa* (right), Cartier Island Marine Park.

Below: The near absence of fast-growing but delicate staghorn and plate corals (*Acropora* species) suggests that Cartier Island reefs were recently damaged by a cyclone or heatwave. A diverse range of other coral species are present. They are patchily distributed over a large area, including the round brain coral *Symphyllia recta* at left.

'*Off* limits to everyone for decades due to it being a practice bombing target during the war, and the subsequent risk of unexploded ordinance, we at last got a glimpse of what it could look like elsewhere if the reefs were afforded the same protection. Abundant corals, larger fish and more diversity – if proof was needed that protection works, Cartier Island is the evidence.'

IAN SHAW, RLS BLOG.

Once night arrives, the reef transforms – invertebrates including soft corals emerge (below), while most fishes disappear. Cartier Island Marine Park.

Above left: Although populations vanished from nearby Ashmore Island Marine Park about 20 years ago, sea snakes remain common in Cartier Island Marine Park. Turtle-headed Sea Snake (*Emydocephalus annulatus*).

Above right: Green Turtles (*Chelonia mydas*) spend nearly half the day sleeping in reef crevices and depressions in the seabed. They can sleep underwater for several hours without returning to the surface for a breath. Cartier Island Marine Park.

Left: Most small marine invertebrates are brilliantly coloured, even those belonging to groups typified by ugly species on land. Glorious Flatworm (*Pseudobiceros gloriosus*), Cartier Island Marine Park.

Kimberley Marine Park

The Kimberley Marine Park includes Wunambal Gaambera, Dambimangari, Mayala, Bardi Jawi and the Nyul Nyul people's Sea Country. Many of these Traditional Owners have had their native title rights recognised in their Sea Country, including in parts of the Kimberley Marine Park managed by Parks Australia. In addition, most marine areas inshore of the Australian Marine Park are managed by the Western Australian Government jointly with Traditional Owners, including co-design of management plans.

Underwater habitats in the region are shaped by massive tides along the southern Kimberley coast, which generate huge tidal currents and turbid conditions.

Zone Name	Zone Area (km²)
National Park Zone (no-fishing)	6,392
Habitat Protection Zone (recreational and limited commercial fishing)	5,665
Multiple Use Zone (recreational and commercial fishing)	62,411

FISH BIOMASS	FISH RICHNESS	CRYPTIC FISH RICHNESSS	INVERTEBRATE RICHNESS
64kg/500m²	34 species/500m²	5 species/100m²	3 species/100m²

SEA SNAKE DENSITY	SHARK DENSITY	CROWN OF THORNS SEASTAR DENSITY	CORAL COVER
0 snakes/500m²	0.23 sharks/500m²	0.04 COTs/100m²	28% live coral

Mermaid Reef Marine Park

The Mermaid Reef Marine Park covers the northernmost of three Indian Ocean reefs that make up the Rowley Shoals, located 300km west of Broome. It abuts the much larger Argo-Rowley Terrace Marine Park, which extends 400km west into abyssal depths. The three Rowley Shoals reefs are amongst the most perfect examples of shelf-edge atolls in the world.

Mermaid Reef was first protected as a Marine National Nature Reserve by the Australian Government in 1991. No-fishing safeguards have been extended to the present, as outlined in the 2018 Management Plan that spans the North-west Marine Parks Network. The other two Rowley Shoals (Clerke Reef and Imperieuse Reefs) are also marine protected areas, but under WA State legislation that allows recreational and charter fishing in most offshore and some lagoonal areas. The reason that the shoals are managed by different governments relates to historical claims of land above sea level. The sand cay on Mermaid Reef is fully submerged at high tide and thus falls under Commonwealth jurisdiction, while the other two reefs have permanent sand cays above the high-water mark. Spring tides can range up to 4m.

The three reefs are similar in size and shape, with enclosed lagoons, small sand cays and steep outer reef edges that drop to over 300m depth within 2km of shore. Due to the steep slope of the atoll margins, upwelling of nutrients promotes plankton production, which attracts numerous pelagic species such as dolphins, tuna, billfish and sharks. The clear water, frequent presence of large pelagic fishes, and diverse range of habitats attracts divers from around the world. Most concentrate their underwater visits on channels through the outer coral barrier and the near-vertical drop-offs rather than the wave-exposed reef crest or silty inner lagoons.

ANDREW GREEN

MERMAID REEF
(-17.11, 119.62)

RLS Sites

0.5 1 2 3 4 5
kilometres

Zone Name	Zone Area (km²)
National Park Zone (no-fishing)	540

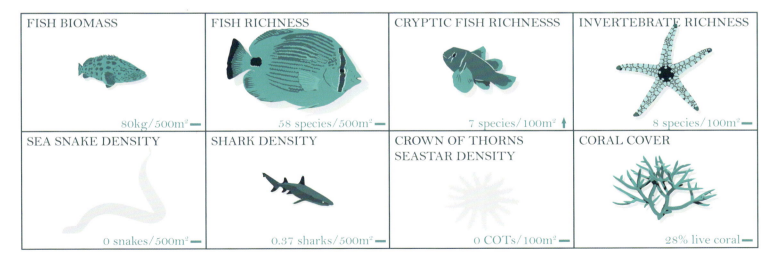

FISH BIOMASS	FISH RICHNESS	CRYPTIC FISH RICHNESSS	INVERTEBRATE RICHNESS
80kg/500m² —	58 species/500m² —	7 species/100m² ↑	8 species/100m² —

SEA SNAKE DENSITY	SHARK DENSITY	CROWN OF THORNS SEASTAR DENSITY	CORAL COVER
0 snakes/500m² —	0.37 sharks/500m² —	0 COTs/100m² —	28% live coral —

Opposite: The reef crest that protects Rowley Shoals atolls from a wave battering comprises a flat pavement that rises above sea level at low tide. Only short stout corals, low-lying sponges, encrusting algae, and the occasional Giant Clam (*Tridacna gigas*) survive. Divers can only visit these locations during periods with unusually calm seas. Imperieuse Reef.

Above: Reef habitat within Rowley Shoals lagoons consist primarily of fragile thickets of *Acropora* staghorn coral interspersed with patches of dead coral rubble. Coral thicket, Clerke Reef lagoon.

Below: Complex reef habitat is located just below the reef crest where corals are bathed in sunlight and in water flowing off the reef top. Mermaid Reef Marine Park.

Above: Currents powered by 3–4m tides surge through channels linking the ocean to inner lagoon. As currents weaken on the turn of the tide, divers exploring these channels are rewarded by a seascape of gorgonians, soft and hard corals, passing schools of pelagic fishes, and groupers. Red sea fan (*Melithaea* species). Mermaid Reef Marine Park.

Left: Lagoon coral thickets provide refuges from predators for small fishes, including for juveniles of reef giants such as Humphead Maori Wrasse (*Cheilinus undulatus*) and Bumphead Parrotfish (*Bolbometopon muricatum*), pictured here. Clerke Reef lagoon.

41

Above: The anemone *Heteractis crispa* provides habitat for a Pink Anemonefish (*Amphiprion perideraion*) in a tidal channel. Mermaid Reef Marine Park.

Below: A wide variety of mobile invertebrate species inhabit channels and outer reefs in this marine park.

Left: Pin Cushion Star (*Culcita novaeguineae*). Mermaid Reef Marine Park. Centre: Tiger Cowry (*Cypraea tigris*). Mermaid Reef Marine Park. Right: Graeffe's Sea Cucumber (*Pearsonothuria graeffei*), Mermaid Reef Marine Park.

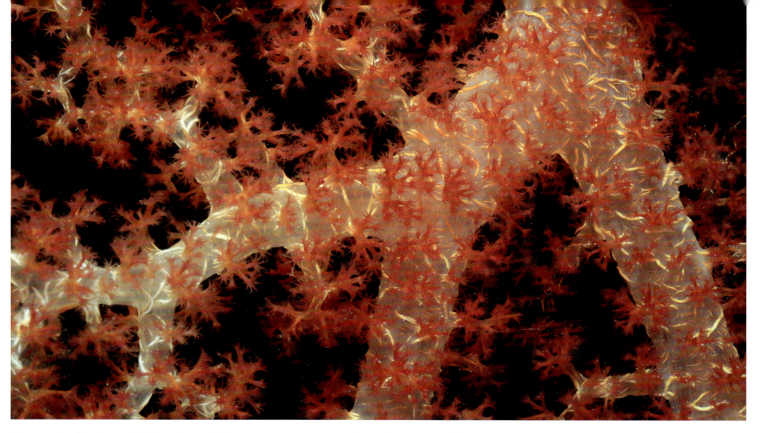

Above: Red soft coral (*Dendronephthya* species). Mermaid Reef Marine Park.

Below: Close inspection of gorgonian fans often reveals invertebrates clinging to the fine branches. Pygmy Seahorses (*Hippocampus bargibanti*) can also be sighted attached to gorgonians in this marine park. Depressed Spider Crab (*Xenocarcinus depressus*). Mermaid Reef Marine Park.

Sand and rubble patches cover large areas on deep outer slopes, thus providing habitat for sand-dwelling species. Bluehead Tilefish (*Hoplolatilus starcki*) hover around burrow entrances. Mermaid Reef Marine Park.

The Redface Squirrelfish (*Sargocentron violaceum*) remains well concealed during daylight, emerging in search of prey at night. Mermaid Reef Marine Park.

Giant Clam (*Tridacna gigas*). Mermaid Reef Marine Park.

Doublesaddle Butterflyfish (*Chaetodon ulietensis*). Mermaid Reef Marine Park.

Lanternfishes and other deep-water species that rarely enter shallows can occasionally be seen at night on the north-eastern outer slopes of Mermaid Reef. Tidal currents funnel the small fishes up narrow canyons from great depth. Lanternfishes are super-abundant in oceanic mid-waters and are a critical food resource for squid, tuna and other pelagic fishes. Different species possess distinctive arrangements of light organs along the sides. Spinycheek Lanternfish (*Benthosema fibulatum*). Mermaid Reef Marine Park.

Above: Clear oceanic waters surrounding the Rowley Shoals support high densities of pelagic fishes, which in turn attract anglers. Sailfish (*Istiophorus platypterus*). North West Shelf.

Below: Specklefin Grouper (*Epinephelus ongus*) at night. Mermaid Reef Marine Park.

Below left: Bigeye Soldierfish (*Myripristis pralinia*), Mermaid Reef Marine Park.

Below right: Checkered Snapper (*Lutjanus decussatus*), Mermaid Reef Marine Park.

Sailfin Snapper (*Symphorichthys spilurus*) at night. Mermaid Reef Marine Park.

Cryptic fishes

Recent studies suggest that the great abundance of large fishes seen on reefs is only possible because of the rapid turnover of the inconspicuous ('cryptic') species hiding amongst cracks[30]. Small cryptic fishes provide a highly productive food source for predators because they are extremely abundant and have a short life cycle. They also produce copious quantities of eggs and larvae, thus supplying planktivorous fish species with food.

Cryptic fish species are also disproportionately important from a conservation perspective because their populations tend to be confined within small regions. Unlike larger fishes in the tropical Australian Marine Parks network that typically extend widely across the Indo-Pacific region, some cryptic fish species are found on a single reef system only. A localised heatwave, cyclone or pollution event could potentially extinguish the species!

Although overlooked by many divers, the variety of colours and patterns in small gobies, blennies and triplefins fascinates those who search them out. The great diversity of these fishes in the Australian Marine Parks network is shown here with examples from just two groups, the *Ecsenius* blennies and the *Eviota* gobies. Many new species await discovery.

Ecsenius lividanalis, Ashmore Reef Marine Park

Left to right: *Ecsenius alleni*, Clerke Reef, Rowley Shoals; *Ecsenius fourmanoiri*, Middleton Reef, Lord Howe Marine Park; *Ecsenius stictus*, Saumarez Reef, Coral Sea Marine Park.

Left to right: *Ecsenius tigris*, Holmes Reef, Coral Sea Marine Park; *Ecsenius yaeyamensis*, Ashmore Reef Marine Park; *Eviota bifasciata*, Ashmore Reef, Coral Sea Marine Park.

Left to right: *Eviota guttata*, Boot Reef, Coral Sea Marine Park; *Eviota prasites*, Ashmore Reef Marine Park; *Eviota sebreei*, Ashmore Reef Marine Park.

Unknown species of *Eviota*, Ashmore Reef Marine Park.   Unknown species of *Eviota*, Mermaid Reef Marine Park

North Marine Parks Network

The North Marine Parks Network extends across a large area of northern Australia, from the Joseph Bonaparte and Oceanic Shoals Marine Parks in the west to the West Cape York Marine Park in the east. It covers a total of 157,480km². Coral reef communities within the network are biogeographically linked in multiple directions, with influences from Papua New Guinea, Torres Strait and northern Great Barrier Reef to the east, and Indonesia and the northern Indian Ocean to the west.

Aboriginal and Torres Strait Islander people continue to assert inherited rights and responsibilities over Sea Country within the North Network. It is recognised that Sea Country extends from terrestrial areas into nearshore and offshore waters; and that songlines traverse Sea Country. Sacred sites are also located in Australian Marine Parks in the North Network. Marine animals are recognised for their spiritual values, and their importance for the health and wellbeing of Aboriginal and Torres Strait Islander communities.

Reefs within the eight North Marine Parks emerge from deeper water as gems scattered across the northern continental shelf. North Marine Parks reefs are located far from urban centres off the coasts of the Northern Territory and Queensland, consequently any journey here is firstly a voyage of discovery. Most reefs have never been visited by divers, so most plants and animals on the reefs have been largely hidden from people, and their diversity unappreciated.

In addition to remoteness, divers visiting the North Network are confronted by powerful and unpredictable currents driven by water flowing into and out of the Gulf of Carpentaria and past Timor. Strong winds prevailing from the east can also stir up sediments and reduce visibility. However, colourful undersea gardens of filter-feeding invertebrates thrive in the currents, overcompensating for these challenges.

The scene most often observed on shallow reefs in the North Network comprises a multi-coloured quilt splashed by pink corals, red gorgonians, apricot seawhips, purple volcano sponges, feathery white hydroids, green and brown turfing algae, and white sand patches. Schools of fusiliers and trevally pass by in diagonal shafts of sunlight. Large grouper are occasionally seen, and are more common than elsewhere; however, sharks are seen less frequently than expected for the remote location, perhaps because of ongoing

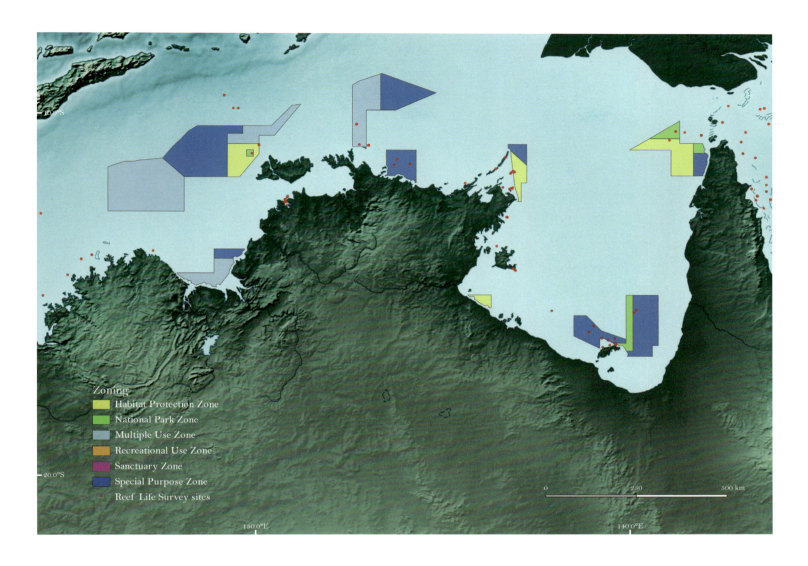

Zoning

- **Habitat Protection Zone**
- **National Park Zone**
- **Multiple Use Zone**
- **Recreational Use Zone**
- **Sanctuary Zone**
- **Special Purpose Zone**
- **Reef Life Survey sites**

fishing pressure when swimming outside no-fishing marine parks. Only two shallow reefs within the North Network are fully protected from fishing – Moss Shoal in the Oceanic Shoals Marine Park and Merkara Reef in the West Cape York Marine Park, both within small National Park Zones. Recreational fishing and some forms of commercial fishing are permitted on all other shallow reefs.

Moving inshore from the more typical North Network reefs, underwater visibility declines and the seabed tends to become sandier, with many of the shallow shoals marked on charts comprising large sand ridges, thus lacking rock or coral. In locations furthest offshore, reefs transition towards clear-water systems more characteristic of Indonesia. Corals are more diverse, generating reef habitat with greater structural intricacy. These complex habitats in turn provide home to many wide-ranging Indo-Pacific reef fishes and invertebrates that are absent from inshore reefs, while fewer endemic Australian species (i.e. species not seen in other countries) are present.

Oceanic Shoals Marine Park

The Oceanic Shoals form a string of shallow pearls dotted along the outer edge of the Northwest Shelf near Indonesian and Timorese territory. They also include a patchwork of shallow reefs and sand ridges scattered across the extensive continental shelf inshore of these outer reefs. The Oceanic Shoals Marine Park encompasses a large proportion of the central shelf reefs but does not extend to the shelf edge. Stopping for a dive in these locations, where the blue horizon can extend in all directions for over 200km, is both unsettling and exhilarating.

Reef ecosystems vary depending on how far offshore they are. Numbers of Indo-Pacific species of corals and fishes increase progressively offshore towards the shelf edge, while mobile invertebrates, turfing algae and sand decrease. Anecdotal stories from divers visiting the region 40 years ago indicate numbers of sharks, groupers, trevally, mackerel, tunas and lobsters that now seem extraordinarily high. Following a decline in stocks of large fishes and sea cucumbers across Indonesia, fishers using traditional vessels increasingly access the region under a Memorandum of Understanding between governments. This may have contributed sufficient fishing pressure to reduce populations of exploited species to levels similar to elsewhere across the Indian Ocean.

Zone Name	Zone Area (km²)
National Park Zone (no-fishing)	406
Habitat Protection Zone (recreational and limited commercial fishing)	6,929
Multiple Use Zone (recreational and commercial fishing)	39,964
Special Purpose Zone (Trawl) (recreational and commercial fishing)	24,444

FISH BIOMASS	FISH RICHNESS	CRYPTIC FISH RICHNESSS	INVERTEBRATE RICHNESS
59kg/500m² —	28 species/500m² —	5 species/100m² —	3 species/100m² —

SEA SNAKE DENSITY	SHARK DENSITY	CROWN OF THORNS SEASTAR DENSITY	CORAL COVER
0 snakes/500m² —	0.38 sharks/500m² —	0.13 COTs/100m² —	26% live coral —

Above: When conditions are suitable for feeding, small white polyps emerge to speckle the branches of the lyre gorgonian (*Ctenocella pectinata*), muting its bright red coloration. Each gorgonian fan comprises a colony of thousands of polyps, the fan orientated perpendicular to water flow to maximise capture of particles drifting past polyp tentacles. Currents tend to flow in one direction through the Oceanic Shoals Marine Park due to strong outflow of water from the Pacific to the Indian Ocean through the geographic gap between Timor and Australia (the 'Indonesian Throughflow'). Plankton-feeding organisms from gorgonians and sponges to whales and Whale Sharks capitalise on these highly productive waters. Marie Shoal, Oceanic Shoals Marine Park.

Below: Most reef habitats in the Oceanic Shoals Marine Park have little vertical structure, and few large crevices exist for cave-dwelling animals. Ornate Spiny Lobsters (*Panulirus ornatus*) compete for the little available space to minimise risk of attack by predatory sharks, rays and grouper. Marie Shoal, Oceanic Shoals Marine Park.

Fragile Sea Whips (*Junceella fragilis*) act as sea vanes on reefs, indicating direction of currents and also strength; the most powerful currents flatten sea whips against the seabed. Marie Shoal, Oceanic Shoals Marine Park.

Above: A mosaic of life crowds onto shallow reefs in the Oceanic Shoals Marine Park, including gorgonians (*Ctenocella pectinata*, left; *Rumphella* sp., bottom left), corals (*Favites* sp., foreground and *Turbinaria* sp. centre), sea whips (*Junceella fragilis*), and green algae (*Halimeda macroloba*, right; *Caulerpa fergusonii*, left). Marie Shoal, Oceanic Shoals Marine Park.

Left: The Raspy Sea Urchin (*Prionocidaris baculosa*) is rarely seen by divers elsewhere, but occurs in dense aggregations on the seabed on Parry Shoal. Urchins affect habitat available for other invertebrates and fishes by scraping the rocky bottom using chisel-shaped teeth, thus clearing the seabed of seaweeds and settling invertebrates. They add microhabitat for animals living amongst spines, such as the hermit crabs seen here. Parry Shoal, Oceanic Shoals Marine Park.

Arafura Marine Park

Arafura Marine Park is the northernmost marine park in Australian waters, reaching 9oS latitude, extending further north than the southern tip of New Guinea. The Yuwurrumu members of the Mandilarri-Ildugij, the Mangalara, the Murran, the Gadura-Minaga and the Ngaynjaharr clans have responsibilities for Sea Country in the Arafura Marine Park. These clans have had native title rights recognised for their Sea Country, including over a part of the Arafura Marine Park.

Environmental conditions vary widely from the near-coastal reefs at Bramble Rocks off Croker Island in the south to clear-water coral reefs at Money Shoal in more offshore waters. Tidal currents can run strongly but tend to be less powerful with distance offshore. The total area of this marine park is 22,925km², all open to recreational and some forms of commercial fishing; trawlers can operate in 46 per cent of the total area.

Coral bommies, Money Shoal, Arafura Marine Park.

Zone Name	Zone Area (km²)
Special Purpose Zone (recreational and commercial fishing)	42
Multiple Use Zone (recreational and commercial fishing)	12,422
Special Purpose Zone (Trawl) (recreational and commercial fishing)	10,461

FISH BIOMASS	FISH RICHNESS	CRYPTIC FISH RICHNESSS	INVERTEBRATE RICHNESS
82kg/500m² —	46 species/500m² —	7 species/100m² —	6 species/100m² —

SEA SNAKE DENSITY	SHARK DENSITY	CROWN OF THORNS SEASTAR DENSITY	CORAL COVER
0 snakes/500m² —	0.03 sharks/500m² —	0 COTs/100m² —	27% live coral —

Above: Flat table-shaped *Acropora* corals add to the three-dimensional structure of Money Shoal, providing plentiful nooks for large fishes to hide under. Arafura Marine Park.

Below: Money Shoal supports an outlying ecosystem within the North Network due to clear oceanic water touching its location 70km north of nearest landfall at Croker Island. A diverse array of coral species form a complex reef system, including numerous branching staghorn (*Acropora*) species. Arafura Marine Park.

Many species of butterflyfish are specialist feeders on living coral polyps. Because of the diversity of corals, this family is well represented amongst the fish fauna at Money Shoal, including this Triangular Butterflyfish (*Chaetodon baronessa*) out at night. Arafura Marine Park.

Fusiliers are a characteristic feature of reefs in the North Network, forming large schools during daylight hours. Goldband Fusilier (*Caesio caerulaurea*) are also amongst the most common fishes sighted by divers at night, drifting randomly above the reef then darting away from torchlight. Money Shoal, Arafura Marine Park.

Below: With nightfall, a different set of predators search the reef for fish and crustacean prey. The Bigfin Reef Squid (*Sepioteuthis lessoniana*) hunts amongst coral rubble. Money Shoal, Arafura Marine Park.

Prickly soft coral (*Dendronephthya* species), Money Shoal, Arafura Marine Park.

Spanish Mackerel (*Scomberomorus commerson*), tuna and other large pelagic fishes patrol the reef margins. With a flick of the tail, they flash downward, scattering flocks of small coral-associated fishes. Money Shoal, Arafura Marine Park.

Below: Bottlenose dolphins frequently accompany boats travelling through Australian Marine Parks. Two bottlenose species are found in tropical waters: the Offshore Bottlenose Dolphin (*Tursiops truncatus*) prefers oceanic waters, while the Indo-Pacific Bottlenose Dolphin (*Tursiops aduncus*; pictured here in the Arafura Marine Park near Croker Island) is a more coastal species.

Arnhem Marine Park

Located in Commonwealth waters directly north of western Arnhem Land, reefs in the Arnhem Marine Park rise from sandy shoals. The coastal Aboriginal people of West Arnhem Land have responsibilities for Sea Country in the Arnhem Marine Park, including respect for songlines that extend from terrestrial areas into nearshore and offshore waters.

Communities of plants and animals living on the seabed are typical of inshore communities distributed more widely across the North Network, with a predominance of filter-feeding animals that thrive in the strong current flow.

Paxie Shoal, Arnhem Marine Park.

Zone Name	Zone Area (km²)
Special Purpose Zone (recreational and commercial fishing)	7,125

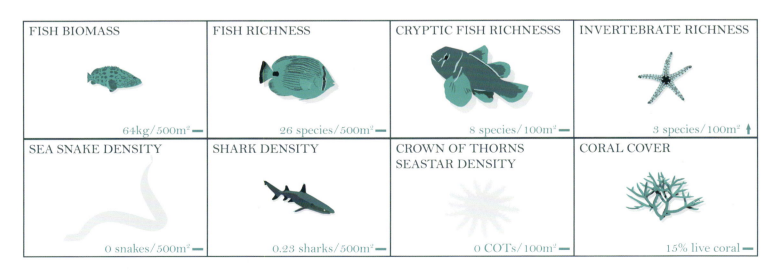

FISH BIOMASS	FISH RICHNESS	CRYPTIC FISH RICHNESSS	INVERTEBRATE RICHNESS
64kg/500m² —	26 species/500m² —	8 species/100m² —	3 species/100m² ↑
SEA SNAKE DENSITY	SHARK DENSITY	CROWN OF THORNS SEASTAR DENSITY	CORAL COVER
0 snakes/500m² —	0.23 sharks/500m² —	0 COTs/100m² —	15% live coral —

Above: Red sea fan (*Melithaea* sp.) with white polyps emergent at night. Gadget Reef, Arnhem Marine Park.

Below: Foliose corals in the genus *Turbinaria* are common on northern reefs. Of the eleven species known worldwide, seven have been recorded during RLS surveys in North Network marine parks. *Turbinaria mesenterina*, Gadget Reef, Arnhem Marine Park.

Left: Most Arnhem Marine Park pinnacles possess a flat invertebrate-covered plateau on top. Gadget Reef, Arnhem Marine Park.

Right: Golden gorgonian with grey polyps. Gadget Reef, Arnhem Marine Park.

Above: Flocks of damselfishes and other plankton feeders living just above the seabed fuel the diets of large predatory fishes on offshore reefs. Sit-and-wait ambush predators such as this Tasselled Wobbegong (*Eucrossorhinus dasypogon*) compete with roving pelagic fishes for the abundant prey. Paxie Shoal, Arnhem Marine Park.

Left: Ambush predators living amongst sponges and corals can possess remarkably effective camouflage, including tassels that break up the outline. During daylight hours they also take advantage of crevices and other irregularities in the seabed surface, concealed from predators, prey and divers in the shadows. Banded Frogfish (*Halophryne diemensis*), Gadget Reef, Arnhem Marine Park.

Right: After nightfall, many mid-sized reef fishes move out from cover, not risking capture by day-feeding sharks. These species include the Doubletooth Soldierfish (*Myripristis hexagona*), a species rarely seen in daylight but frequently observed flitting in the open at night. Gadget Reef, Arnhem Marine Park.

Above: In addition to camouflage, cryptic predators often possess bright red coloration when lit up by camera flash, but appear dull brown in natural light due to rapid absorbance of red light in the first few metres underwater. Blacktip Rockcod (*Epinephelus fasciatus*), Gadget Reef, Arnhem Marine Park.

Below: Characteristic seascape that greets divers on inshore reefs across the North Marine Park network. Foliose corals (*Turbinaria* species), seawhips, soft corals and gorgonians dominate the seabed. Gadget Reef, Arnhem Marine Park.

Above: Pink rope gorgonian (*Rumphella* sp.) with grey polyps, Paxie Shoal, Arnhem Marine Park.

Below: Paxie Shoal, Arnhem Marine Park.

Above: Flocks of damselfishes (*Neopomacentrus cyanomos* and *Neopomacentrus bankieri*) forage on plankton in the open, darting under coral (here *Turbinaria fondens*) and sponges when approached by predators. Paxie Shoal, Arnhem Marine Park.

Below: Barrel Sponges (*Xestospongia testudinaria*) can grow to 2m in height. They provide homes to crustaceans and other small invertebrates that live in internal passages. Seawater pumped through the passages carries an array of planktonic food particles. Paxie Shoal, Arnhem Marine Park.

Wessel Marine Park

The Yolŋu people have responsibilities for Sea Country in the Wessel Marine Park, which extends out from terrestrial areas, and is traversed by songlines.

Massive tidal flows oscillate through the Wessel Marine Park on spring tides, driven by the passage of water entering the Gulf of Carpentaria through gaps between the Wessel Islands. Shallow reefs accessible to divers are largely confined to shoals east of Truant Island, where Crown-of-thorns sea stars and urchins both occur in high numbers. Most reef locations in the Wessel Marine Park remain unexplored.

Zone Name	Zone Area (km²)
Habitat Protection Zone (recreational and limited commercial fishing)	3,811
Special Purpose Zone (Trawl) (recreational and commercial fishing)	2,097

FISH BIOMASS	FISH RICHNESS	CRYPTIC FISH RICHNESSS	INVERTEBRATE RICHNESS
99kg/500m² —	31 species/500m² —	8 species/100m² —	5 species/100m² —

SEA SNAKE DENSITY	SHARK DENSITY	CROWN OF THORNS SEASTAR DENSITY	CORAL COVER
0 snakes/500m² —	0 sharks/500m² —	0.50 COTs/100m² —	30% live coral —

Crown-of-thorns sea star

Crown-of-thorns sea stars (COTs) are typically associated with the Great Barrier Reef, where an active suppression campaign is underway. This involves five boats with dedicated divers injecting individual COTs with bile salts. Nevertheless, RLS divers recorded higher COTs densities in the Wessel Marine Park than during surveys of locations across the length of the Great Barrier Reef. Most corals in the Wessel Marine Park possessed feeding scars. This finding was unexpected given that the main hypotheses proposed for COTs outbreaks involve overfishing of predators or poor water quality, both of which are considered low in remote waters off the Top End. Although predatory fishes are more abundant in the Wessel Marine Park than in most other marine parks where selective fishing is allowed, local predator populations have perhaps been reduced below the level needed to control COTs outbreaks.

COTs are readily distinguished from other sea stars by their armoury of sharp poisonous spines and grey-green coloration with additional orange-brown markings. During outbreaks – as observed in the Wessel Marine Park surveys – COTs remain in the open and consume coral both day and night, whereas they remain hidden in crevices during the day when in low numbers.

This sea star has long been known scientifically as *Acanthaster planci*, but recent molecular evidence indicates that at least four species exist in different parts of its Indo-Pacific range. True *Acanthaster planci* possess iridescent purple coloration and are restricted to the northern Indian Ocean. The correct name for the Australian species is probably *Acanthaster solaris*, but this remains to be confirmed. Gadget Reef, Arnhem Marine Park.

Gulf of Carpentaria Marine Park

The Gulf of Carpentaria is best visualised as a large (350,000km²) internal sea with a flattish seabed of silt and sand, and with seagrass and mangroves around the rim. In the south, the Lardil people of the Wellesley Islands have had their native title rights recognised in their Sea Country, including in part of the Gulf of Carpentaria Marine Park. They maintain a continuing spiritual connection with Sea Country, and cultural responsibilities for its management.

Large stocks of prawns support a substantial fleet of prawn trawlers. Shallow ridges with patchy coral cover were first reported in the southern gulf in 2004, covering an area of about 80km². These submerged hilltops are considered remnants of reefs that were much more active thousands of years ago, when the climate was cooler and sea level lower – a time when environmental conditions were more suitable for coral growth. Boundaries of the Gulf of Carpentaria Marine Park were designed to encompass most of these ancient coral systems, albeit with prawn trawling allowed across most of the park, including the shallow reef habitats. Tidal currents are much weaker than elsewhere in the North Network, rarely exceeding 1m change, and with a single tidal cycle each day.

Zone Name	Zone Area (km²)
National Park Zone (no-fishing)	3,623
Special Purpose Zone (Trawl) (recreational and commercial fishing)	20,148

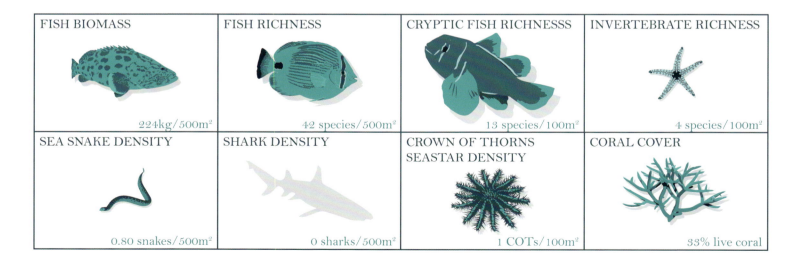

FISH BIOMASS	FISH RICHNESS	CRYPTIC FISH RICHNESSS	INVERTEBRATE RICHNESS
224kg/500m²	42 species/500m²	13 species/100m²	4 species/100m²

SEA SNAKE DENSITY	SHARK DENSITY	CROWN OF THORNS SEASTAR DENSITY	CORAL COVER
0.80 snakes/500m²	0 sharks/500m²	1 COTs/100m²	33% live coral

Above: Sea snakes were observed more frequently underwater in the Gulf of Carpentaria Marine Park than in other Australian Marine Parks. Few dives would be completed without at least one snake following divers. Olive Sea Snakes (*Aipysurus laevis*) observed in the gulf are very different in appearance to the north-eastern and north-western Australian populations, and are possibly a distinct species. Carpentaria Reef, Gulf of Carpentaria Marine Park.

Left: Shallow seamounts in the Gulf of Carpentaria Marine Park often have a flat upper plateau with large patches of sand, while the current-swept upper slopes support denser communities of filter-feeding animals. Carpentaria Reef, Gulf of Carpentaria Marine Park.

Right: The sponge community in northern Australian waters is extremely diverse but with most species still unnamed. The Ambon Tube Sponge (*Haliclona amboinensis*) is an exception due its conspicuous form and prevalence. Chinamans Reef, Gulf of Carpentaria Marine Park.

Above: Tangle of gorgonians. Chinamans Reef, Gulf of Carpentaria Marine Park.

Left: A Lyre Gorgonian (*Ctenocella pectinata*) provides cover for small fishes. Chinamans Reef, Gulf of Carpentaria Marine Park.

Below: Lyre Gorgonian (*Ctenocella pectinata*), Chinamans Reef, Gulf of Carpentaria Marine Park.

Above: A diverse array of bottom living fishes hide amongst the dense jumble of invertebrates attached to the seabed. Cryptic fishes range in size from small 2.5cm gobies to 50cm groupers. Yellowspotted Rockcod (*Epinephelus areolatus*). Chinamans Reef, Gulf of Carpentaria Marine Park.

Below: Seamounts in the southern Gulf of Carpentaria are sufficiently remote for species overfished elsewhere to persist in moderate numbers. Queensland Grouper (*Epinephelus lanceolatus*). Groper Shoal.

West Cape York Marine Park

West Cape York Marine Park abuts the western entrance to Torres Strait, lying across a major pathway for ships transiting between the Pacific and Indian Oceans. As elsewhere in the North Network, currents during the ebb and flood of spring tides are extremely strong, in this case due to a 3m tidal range and seawater oscillating into and out of Torres Strait.

Aboriginal and Torres Strait Islander people continue to assert inherited rights and responsibilities over Sea Country within West Cape York Marine Park.

Zone Name	Zone Area (km²)
National Park Zone (no-fishing)	3,329
Habitat Protection Zone (recreational and limited commercial fishing)	10,114
Special Purpose Zone (recreational and commercial fishing)	2,569

FISH BIOMASS	FISH RICHNESS	CRYPTIC FISH RICHNESSS	INVERTEBRATE RICHNESS
64kg/500m² —	24 species/500m² —	4 species/100m² —	3 species/100m² —
SEA SNAKE DENSITY	SHARK DENSITY	CROWN OF THORNS SEASTAR DENSITY	CORAL COVER
0.29 snakes/500m² —	0.52 sharks/500m² —	0 COTs/100m² —	17% live coral ↓

Above: RLS investigations revealed major ecosystem changes over the past decade within the West Cape York Marine Park. During initial baseline surveys in 2012, divers discovered diverse assemblages of attached invertebrates such as these pink-pigmented soft corals (left), *Chromonephthea* species, Carpentaria Shoal. However, fine turf seaweeds and the brown alga *Padina* now blanket the seabed, and sponges overgrow corals and gorgonians (right), Merkara Shoal. The cause of this habitat degradation remains debatable, but most likely relates to a major heatwave that swept through the Gulf of Carpentaria in early 2016, also killing over 7,000 hectares of coastal mangroves[31].

Above left: The pearl perch family includes only four species that are all mostly found in Australian waters. The Threadfin Pearl Perch (*Glaucosoma magnificum*) has a range that extends to southern New Guinea, and its attractiveness draws the attention of divers in the North Marine Parks Network. The pictured individuals were observed just outside the West Cape York Marine Park at Proudfoot Shoal. To understand if ecological changes in Australian Marine Parks through time relate primarily to park management or to changing climate and other broad-scale effects, RLS divers gain broader context by monitoring nearby reefs outside the park boundaries.

Above right: The Flatback Turtle (*Natator depressus*) – one of only seven sea turtle species worldwide – is restricted to northern Australia. This region also supports feeding and breeding grounds for four of the other six sea turtle species. Crab Island, located just east of West Cape York Marine Park, includes important nesting beaches for Flatback Turtle. Juveniles hatch and disperse westward to forage on marine park shoals. Crab Island.

Coral Sea Marine Park

The Coral Sea Marine Park is Australia's largest marine park, encompassing an area three times the size of the Great Barrier Reef Marine Park[32,33]. It covers offshore waters south from Cape York Peninsula to just north of the latitude of Bundaberg as a multi-zoned park. The Great Barrier Reef lies along the park's western border while along its other margins sit New Caledonia, Papua New Guinea, the Solomon Islands, and the Tasman Sea. Scattered across this region are approximately 34 reefs and 56 cays and islets. The Coral Sea's remote location and low level of human impact make it one of the last 'pristine' seas in the world[34].

The Coral Sea Marine Park has a diverse underwater landscape of seamounts, canyons, shallow tropical ecosystems, and deep abyssal plains extending nearly 5,000m below the ocean surface. These features support a rich diversity of life including over 200 coral and 700 fish species. The Coral Sea's marine life is distinct from the neighbouring Great Barrier Reef, with closer similarity to Pacific islands such as Tonga and Samoa that are over 2,500km away[33].

Distinct ecological communities are also distributed within the park, which is influenced by a complex, changing system of ocean currents. Reef communities subdivide across the centre of the Coral Sea Marine Park due to the South Equatorial Current, which splits into two, one branch flowing north (the Hiri Current), the other south (East Australian Current). This division creates a barrier to species mixing, which, combined with the low input from the surrounding areas, makes the Coral Sea a largely self-sustaining environment. This increases the risk of local extinctions and overexploitation on Coral Sea reefs, and makes it disproportionately vulnerable to climate change and cyclones.

Torres Strait Islanders and coastal Aboriginal people of Cape York continue to assert inherited rights and responsibilities over Sea Country within and adjacent to the Coral Sea Marine Park. Some reefs have been managed by Traditional Owners for thousands of years. The use of natural resource forms part of traditional culture, and is entwined with spirituality. The Meriam People's Sea Country extends over the Ashmore Reef region of the Coral Sea Marine Park. Under traditional (Malo's) law, Ashmore Reef is a significant cultural area. An Indigenous turtle fishery operates in this area using hand collection and

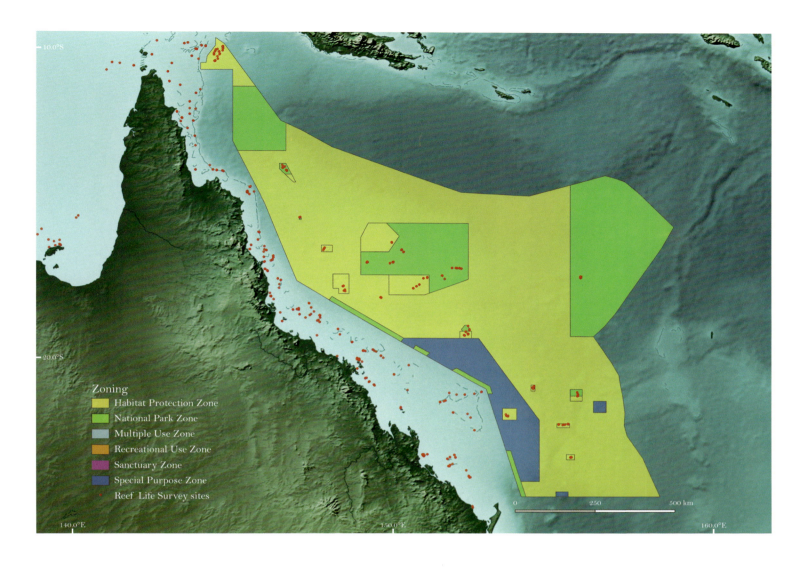

traditional spear methods. More recent human heritage also exists in the Coral Sea Marine Park, with at least 45 historic shipwrecks, dating back to at least the early 1800s.

Visiting the Coral Sea Marine Park offers a once-in-a-lifetime experience. The northern areas have the richest fish and coral diversity, while in the south divers marvel at the profuse coral cover and abundance of sea snakes. Six of the world's seven sea turtle species are found here. It is also a migratory pathway for Humpback Whales.

While located far offshore, many opportunities exist for diving, snorkelling, nature watching, fishing and boating, especially near reefs and atolls. Osprey Reef in the north is one of the world's best dive sites, known for its spectacular coral walls and abundance of reef sharks. Nearby, Black Marlin aggregate to spawn. Travelling south you could meet Potato Cod at Bougainville Reef or drift over colourful plate corals at Holmes Reef. Those who venture further offshore to Mellish Reef may sight ancient *Porites* coral bommies that span up to 19m across – a size estimated to take over 500 years to reach.

In the centre of the park, 300km offshore, are the spectacular Coringa-Herald and Lihou reefs and

Cirrhitichthys falco, Boot Reef.

cays. These reefs have been protected from fishing as national nature reserves since 1982, decades before the current Coral Sea Marine Park was enacted. Twenty-four islets and cays are sprinkled across the area forming an internationally listed wetlands (Ramsar) site with coral reefs in near pristine condition. These uninhabited and rarely visited islands provide an idyllic sanctuary for birds and sea turtles to nest. On the islands are forests of the globally rare *Pisonia grandis*, a flowering tree that provides shelter for nesting birds.

	Zone Name	Zone Area (km²)
Coral Sea Marine Park	*National Park Zone (no-fishing)*	238,400
Coral Sea Marine Park	*Habitat Protection Zone (recreational and limited commercial fishing)*	684,956
Coral Sea Marine Park	*Special Purpose Zone (Trawl) (recreational and commercial fishing)*	66,480

Boot Reef

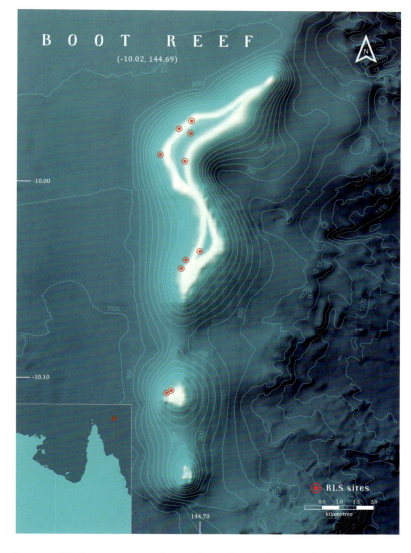

Boot Reef – the most northerly reef in the Coral Sea Marine Park – is positioned 15km north-east of Ashmore Reef, and only 5km from the maritime border with New Guinea. From a diver's perspective, it is also the most fascinating, rewarding anyone who manages to overcome the many challenges of reaching and stopping in this remote location. Fish and coral diversity are exceptionally high compared to other reefs in Australian Marine Parks, as is the range of habitat types.

The reef is S-shaped, each of the two branches fully enclosing a sheltered lagoon with intricate coral gardens. Sheer outer walls drop vertically on all sides, affording no anchorage for visiting boats unless they can enter the lagoon over the reef on a spring tide or hang off from an anchor placed on the reef edge. Most dives down the reef drop-offs involve visits by curious Grey Reef, Whitetip and Silvertip Sharks, which are often keener to closely inspect intruding divers than the other way around.

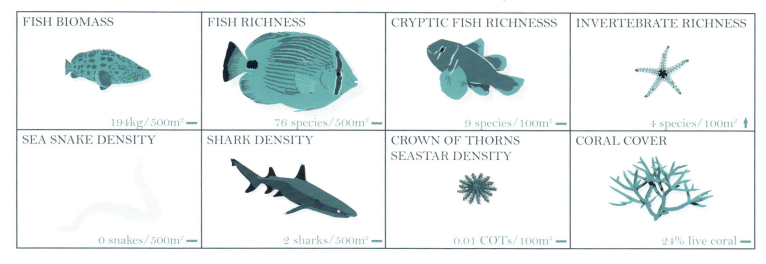

FISH BIOMASS	FISH RICHNESS	CRYPTIC FISH RICHNESSS	INVERTEBRATE RICHNESS
194kg/500m² —	76 species/500m² —	9 species/100m² —	4 species/100m² ↑
SEA SNAKE DENSITY	SHARK DENSITY	CROWN OF THORNS SEASTAR DENSITY	CORAL COVER
0 snakes/500m² —	2 sharks/500m² —	0.01 COTs/100m² —	24% live coral —

'From Ashmore we went even further north to Boot Reef, almost on the Papua New Guinea border. This is the remotest of the remote, virtually unknown to anyone but a few specialists. Our approach was in keeping with the best traditions of the early navigators. As there is no passage into the lagoon, team leader Graham got into the water to make sure there was enough water for *Eviota* to clear the corals clawing the surface. With additional lookouts posted on both bows, we glided across the top of the reef at high tide to seek a sheltered and safe place to anchor.'

BILL BARKER, RLS BLOG.

'Boot Reef emerges rapidly from depth (900–1,500m), creating amazing wall diving. On my first dive, we were greeted by a little Grey Reef Shark, checking us out as we jumped in. The reef wall was beautiful with lovely corals and some huge gorgonians. Fish life was incredibly abundant and biodiversity was high. There were almost an overwhelming number of fish to try and count… three Silvertip Sharks, three Grey Reef Sharks and to top off our shark count, a Leopard Shark (Stegostoma fasciatum)! My second dive had a huge school of Bumphead Parrotfish (Bolbometopon muricatum), probably 50 strong.'

KIRSTY WHITMAN, RLS BLOG.

Giant Trevally (*Caranx ignobilis*) spawning aggregation, Boot Reef drop-off.

'*Boot Reef was a wild and spectacular culmination to our journey.*'

BILL BARKER, RLS BLOG.

Above left: Ringtail Unicornfish (*Naso annulatus*), Boot Reef drop-off.

Above right: Golden Damsel (*Amblyglyphidodon aureus*), Boot Reef drop-off.

Below left: Drummers contribute to the health of reefs by browsing algae and consuming drift material. Snubnose Drummer (*Kyphosus cinerascens*). Boot Reef.

Below right: Flounder match their patterning to the substrate below. The match is mostly – but not always – perfect. Flowery Flounder (*Bothus mancus*) on a coral bommie. Boot Reef.

Several tropical species that are rare elsewhere in Australian waters thrive here, in close proximity to New Guinea. Striped Boxfish (*Ostracion solorensis*), Boot Reef.

The Giant Sea Urchin (*Chondrocidaris gigantea*) has not been recorded in Australian waters outside Boot Reef lagoon.

Above left: Close searches reveal numerous small cryptic fishes (such as the spongegoby *Pleurosicya micheli*) blending amongst sponges and corals. Boot Reef.

Above right: Invertebrates hide in reef cracks. Hairy-netted Hermit Crab (*Aniculus retipes*). Boot Reef.

Below: Most corals fluoresce brightly when illuminated using an ultraviolet ('black light') torch at night. Boot Reef.

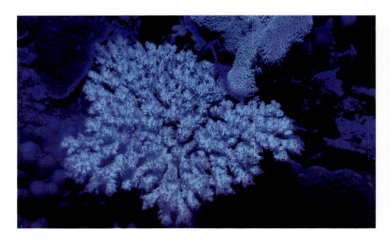

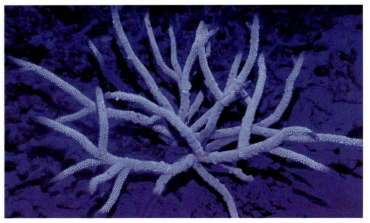

Silvertip Shark (*Carcharhinus albimarginatus*), Boot Reef.

'On this occasion, things were a bit too hectic. The visibility wasn't outstanding, about 20m, and the sharks just wouldn't go away. The five or so Grey Reef Sharks who kept checking on us were quickly joined by two Silvertips. We were approaching the end of our first transect when things suddenly became a lot more complicated. A huge school of surgeonfish rocketed up the wall, closely pursued by 15 or so Grey Reef Sharks swimming up, down and around us at great speed. I've witnessed controlled shark feeds, but being in the water with sharks in large numbers doing what they do naturally is a completely different experience.'

BILL BARKER, RLS BLOG.

Ashmore Reef

Not to be confused with Ashmore Reef off north-western Australia, a second Ashmore Reef rises from deep water as a large atoll in the northern Coral Sea[33]. It is located only 40km east of northernmost tip of the Great Barrier Reef but the two reef systems are quite distinct, separated by water depths exceeding 600m. A thin strip of reef perforated by numerous narrow channels forms the perimeter of the atoll, protecting a deep (to 70m) lagoon within, which extends 40km north to south, 15km east to west. A flat silty plain underlays the centre of the lagoon. Much more interesting to divers than this central basin is the scattering of vertical-walled bommies that reach up to sea level from depths of 20–30m along the eastern lagoon margin. Each contains intricate undercuts and caves, with large gorgonians and cruising pelagic fishes. The exposed eastern face of the atoll is rarely if ever dived because waves driven by the prevailing south-easterly winds crash onto the shore almost continuously. Dense schools of pelagic fishes are associated with the vertical drop-offs that disappear downwards out of sight along the north-west margin. Ashmore Reef has been less affected by cyclones and heatwaves than reefs in the central Coral Sea, so has much better coral cover.

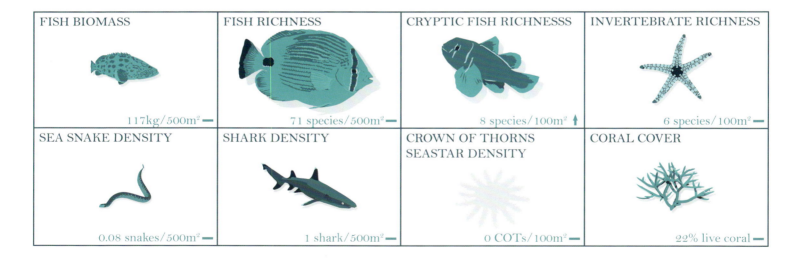

FISH BIOMASS	FISH RICHNESS	CRYPTIC FISH RICHNESSS	INVERTEBRATE RICHNESS
117kg/500m² —	71 species/500m² —	8 species/100m² ↑	6 species/100m² —
SEA SNAKE DENSITY	SHARK DENSITY	CROWN OF THORNS SEASTAR DENSITY	CORAL COVER
0.08 snakes/500m² —	1 shark/500m² —	0 COTs/100m² —	22% live coral —

Ashmore Reef's outer barrier meanders over the horizon.

ASHMORE REEF
CORAL SEA
(-10.26, 144.48)

● RLS sites

kilometres

IAN SHAW

Above: The outer reef drop-off of Ashmore Reef falls near vertically to over 600m depth, with many undercuts. Red sea fan (*Melithaea* species).

Left: Although rarely noticed, most sea whips on vertical walls will host one or two small gobies. Seawhip Goby (*Bryaninops yongei*). Ashmore Reef.

Below: Delicate lace corals guard crevice entrances. *Stylaster* species. Ashmore Reef.

Left: Filter-feeding invertebrates aggregate on current-swept outer walls and the narrow passes into the lagoon. Bennett's Feather Star (*Anneissia bennetti*). Ashmore Reef.

Right: Filter-feeding ascidian. Ashmore Reef.

Below: Zoanthids resemble small anemones but are colonial species connected at the base. This unnamed species occurs in dense colonies at Ashmore Reef. *Palythoa* species.

IAN SHAW

Swallowtail Angelfish (*Genicanthus melanospilos*) feed on plankton on deep reefs. They are unusually common on western drop-offs at Ashmore Reef. The two sexes have strikingly different colour patterns. Female. Ashmore Reef.

Male. Ashmore Reef.

Yellowspeckled Puller (*Chromis alpha*) are specialised for life capturing plankton off verticals walls at depths greater than 12m. Ashmore Reef.

Male Purple Queen (*Pseudanthias tuka*). Ashmore Reef.

IAN SHAW

Above: Complex reef structure on drop-offs includes multi-dimensional crevices, allowing good fits for fishes of a great variety of body size. Slingjaw Wrasse (*Epibulus insidiator*). Ashmore Reef.

Above: Spinyhead Cardinalfish (*Pristiapogon kallopterus*). Ashmore Reef.

Above left: Foursaddle Grouper (*Epinephelus spilotoceps*) living on inner reef margin. Ashmore Reef.

Above right: Crowned Turbinaria (*Turbinaria ornata*) can be found growing on reef tops, where its compact algal form survives wave action and its sharp-edged branches inhibit browsing by fishes. Ashmore Reef.

Above: A series of narrow crevices in a 'spur-and-groove' arrangement efficiently sheds the massive quantities of water crashing onto the reef edge. Ashmore Reef.

Below: Olive Sea Snakes (*Aipysurus laevis*) are abundant in Ashmore Reef lagoon, where they forage for small fishes.

Above: Sunlight falling on shallow reefs generates high productivity of turf algae, which in turn fuel a crowded food web of grazers such as this school of Convict Surgeonfish (*Acanthurus triostegus*), and their predators. Ashmore Reef.

Below: A wide diversity of small cryptic fishes is associated with lagoon bommies.
Left: Largemouth Threefin (*Ucla xenogrammus*). Ashmore Reef. Right: Whitelined Rockcod (*Anyperodon leucogrammicus*). Ashmore Reef lagoon.

93

IAN SHAW

Above left: Bumphead Parrotfish (*Bolbometopon muricatum*) play a unique role on reefs by consuming massive quantities of coral then transforming it through their gut into calcium carbonate sand. This assists development of sand cays. Schools of 20–100 individuals move about in noisy schools, crunching reef coral. Although globally threatened with populations that have disappeared through much of its range, this parrotfish remains common at Ashmore Reef.

Above right: Massive corals form the inner reef margin at Ashmore Lagoon.

Below: Thickets of staghorn coral interspersed with white sand occur in sheltered shallows of Ashmore Lagoon.

Above: Dusky Gregory (*Stegastes nigricans*) tend and aggressively defend small gardens of turf algae within staghorn coral thickets, but active defence by a single damselfish has little hope when large mobs of browsing fishes pass through. Ashmore Reef.

Above: Parrotfishes are the predominant herbivores on most coral reefs, with a critical ecological role. Some species scrape fine turf algae from reef surfaces, generating space for newly settling corals. Others ingest coral rock with associated algae, pooping out plumes of fine calcareous particles that contribute to reef cement. Sixband Parrotfish (*Scarus frenatus*). Ashmore Reef.

Above: Broadclub Cuttlefish (*Sepia latimanus*) flash intruders with everchanging patterns and coloration. Females lay eggs in branching corals within the lagoon. Ashmore Reef.

Below: Planktivorous cardinalfishes find shelter amongst corals in Ashmore Reef lagoon. Girdled Cardinalfish (*Taeniamia zosterophora*).

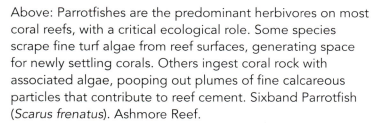

Above left: The Lunartail Bigeye (*Priacanthus hamrur*) is amongst the largest of predatory fishes that permanently reside amongst staghorn corals. Ashmore Reef.

Above right: Noah's Giant Clam (*Tridacna noae*) highlights the poor state of knowledge of coral reef animals. Although conspicuous, it was not recognised as a distinct species until 2014. Ashmore Reef.

Below left: Orangefin Anemonefish (*Amphiprion chrysopterus*) are commonly resident on host anemones on lagoon bommies. Ashmore Reef.

Below right: Black Boxfish (*Ostracion meleagris*). Ashmore Reef.

Above: Ashmore Reef has enormous historical significance as a graveyard for sailing ships. A combination of close proximity to the eastern entrance to Torres Strait, poor charts, and strong currents driven by tides (typically with a vertical range around 3m), resulted in numerous ships running aground during the 1800s. The locations of over 30 of these wrecks still remain unknown. Anchor (left) and cannon (bottom) from the wreck of the 'Comet', a 314-tonne brigantine that ran aground on north Ashmore Reef in 1829.

Below: Currents striking Ashmore atoll rise to the sea surface with associated nutrients, creating an oasis of productivity within the tropical sea. Predatory fishes and large marine mammals concentrate in these areas to capitalise on abundant plankton and dense schools of plankton-feeding fishes. False Killer Whale (*Pseudorca crassidens*). Coral Sea Marine Park near Ashmore Reef.

Osprey Reef

With a tear- (or more accurately planarian-) shape, Osprey is a large 25km-long oceanic atoll that rises vertically upward from the bathyal plain more than 1,000m below. Unlike Ashmore Reef, the narrow perimeter reef is unbroken other than near the entrance channel. Lagoon depths are generally 20–30m in the central basin, with a number of bommies in the south-eastern corner rising to near the surface. Water flow within the lagoon is largely driven by waves crossing the reef rather than through channels, meaning that bommies have less water flow, more algal growth, and fewer pelagic fishes than in better flushed atolls. Undercut caves are common. Some thickets of staghorn coral (*Acropora* species) are present on the lagoon floor, but this habitat was at the epicentre of the catastrophic 2016 heatwave, and only a few living branch tips and remnant patches currently remain. This heatwave also caused extensive bleaching and loss of coral around the atoll margins.

Osprey is world famous as a dive site, primarily because of the wall diving – the seabed disappearing downwards into the deep blue, and because of the high abundances of sharks and other large pelagic species. Sharks concentrate in exceptionally large numbers at the northern tip of the island (North Horn), attracting regular visits by dive charter boats. Recreational fishing at Osprey Reef was banned in 2018, but allowed to continue in nearby Vema and Shark Reefs to the south.

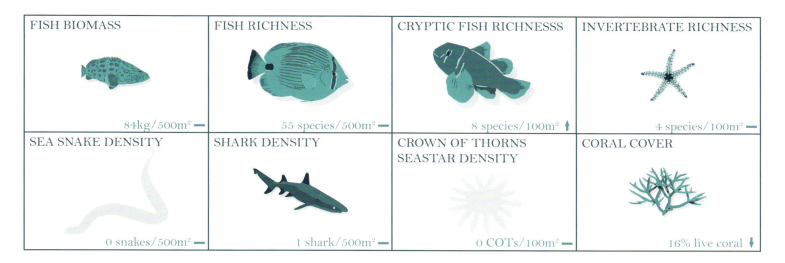

FISH BIOMASS	FISH RICHNESS	CRYPTIC FISH RICHNESSS	INVERTEBRATE RICHNESS
84kg/500m² —	55 species/500m² —	8 species/100m² ↑	4 species/100m² —
SEA SNAKE DENSITY	SHARK DENSITY	CROWN OF THORNS SEASTAR DENSITY	CORAL COVER
0 snakes/500m² —	1 shark/500m² —	0 COTs/100m² —	16% live coral ↓

Top: Sediment accumulation on oceanic atolls is largely dependent on the continuous production of algae in the genus *Halimeda*. These green algae produce necklace-like branches composed of strings of calcareous segments, which eventually die and fall to the seabed. Osprey Reef.

Bottom: Doublebar Goatfish (*Parupeneus crassilabris*) watching passing divers. North Horn, Osprey Reef.

Top: Whitetip Soldierfish (*Myripristis vittata*) is arguably the most characteristic species of deep vertical walls throughout the Coral Sea. They move little, sitting within cave entrances during the day, later becoming active at night. Osprey Reef.

Bottom: The abstract patterning of the Saddle Butterflyfish (*Chaetodon ephippium*) has been labelled the apex of animal coloration. Osprey Reef.

Neon Fusilier (*Pterocaesio tile*), Osprey Reef.

'… *an oasis for living creatures of all kinds.*'

SIR DAVID ATTENBOROUGH.

TONI COOPER

Although fusilier species are nearly always present on coral reefs, continuously swimming past divers in schools, they attract little attention. Nevertheless, they provide over half of the total weight of fishes on many reefs, and play critical ecological roles as consumers of plankton and as prey for large predatory fishes. Osprey Reef.

IAN SHAW

Pearly Nautilus (*Nautilus pompilius*) are rightfully regarded as 'living fossils', the remnant of a very large group of molluscs that persisted with little change over 500 million years. Today, the six nautilus species all live on deep reefs. A sizable population exists around the deeper margins of Osprey Reef, where they hide at around 350m depth during the day, then move up to feed at 100–200m depth at night. Dive charter operators sometimes set baited traps in deep water to capture and showcase these amazing animals.

101

TONI COOPER

To maximise water flow over the arms, feather stars typically occupy locations as high as possible above the seabed. Robust Feather Star (*Himerometra robustipinna*) (resting with arms furled, above), Bennett's Feather Star (*Anneissia bennetti*) (feeding with arms unfurled, below). Osprey Reef.

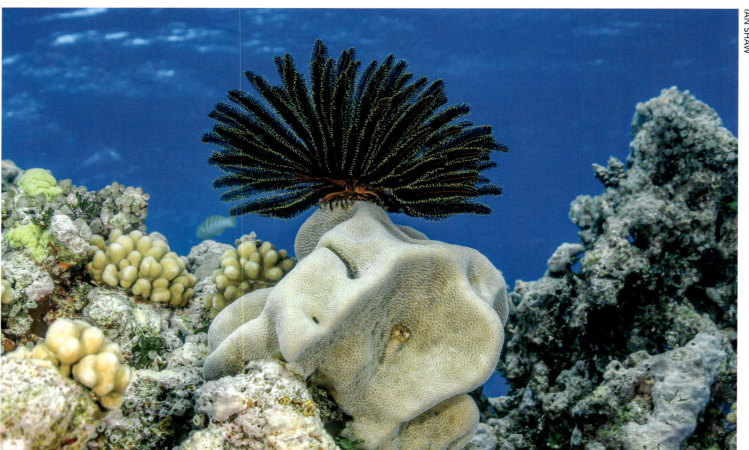

IAN SHAW

Following catastrophic coral bleaching during the early 2016 heatwave, coral surfaces became coated with filamentous algae. The dead corals continue to provide structural habitat for fishes, but will eventually disintegrate. Osprey Reef.

Sailfin Tang (*Zebrasoma velifer*). Osprey Reef.

Like many of their close relatives amongst the butterflyfishes, the two sexes of Longfin Bannerfish (*Heniochus acuminatus*) often swim together as a pair. Osprey Reef.

Bougainville Reef

Bougainville Reef rises as a small atoll from 500m depth. The shallow (less than 5m depth) central lagoon extends about 2km across and is surrounded by reef with no boat channel. The sheer outer walls drop vertically, attracting both large pelagic fishes and divers. Despite isolation from other reef systems, Bougainville is visited by dive charter boats more often than most Coral Sea reefs due to its proximity to Cooktown (190km away) and position *en route* between Cairns and Osprey Reef.

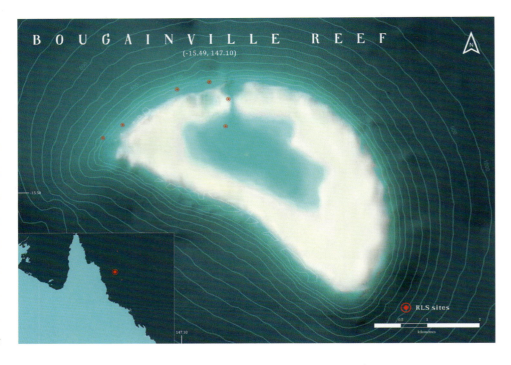

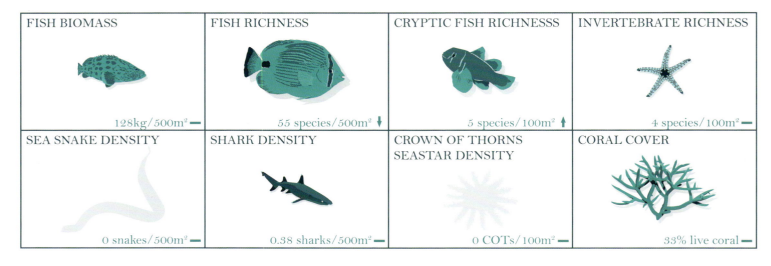

FISH BIOMASS	FISH RICHNESS	CRYPTIC FISH RICHNESSS	INVERTEBRATE RICHNESS
128kg/500m² —	55 species/500m² ↓	5 species/100m² ↑	4 species/100m² —

SEA SNAKE DENSITY	SHARK DENSITY	CROWN OF THORNS SEASTAR DENSITY	CORAL COVER
0 snakes/500m² —	0.38 sharks/500m² —	0 COTs/100m² —	33% live coral —

'An overnight passage then took us to Bougainville Reef, named after the French navigator, Louis-Antoine de Bougainville who, in one of history's great 'what ifs?', was turned away by the breakers of the reef during his search for the Great South Land in 1768. Rising abruptly from the ocean bed almost 2km below, the reef is a dramatic place to dive. We emerged from the water exhilarated by our experience, the sheer fall of the face of the wall and the multitude of fish that swirled around us.'

BILL BARKER, RLS BLOG.

Above left: As is also the case with other Coral Sea atolls, vertical walls surrounding Bougainville Reef attract a wide diversity of large fishes. Snubnose Dart (*Trachinotus blochii*).

Above right: The Mauritian Sea Cucumber (*Actinopyga mauritiana*) – a commercially exploited species that is threatened worldwide – is commonly seen in shallow crevices in locations with good water flow. Bougainville Reef.

Below left: Lemon sponge, *Leucetta chagosensis*. Bougainville Reef.

Below right: Forceps Fish (*Forcipiger flavissimus*) pick individual coral polyps from their calcareous homes with surgeon-like precision. Bougainville Reef.

Above left: The Giant Moray (*Gymnothorax javanicus*) is the largest species of moray, and also the most commonly seen by divers on reefs across Australian Marine Parks. Bougainville Reef.

Above right: Tailspot Lizardfish (*Synodus jaculum*) are amongst the most ferocious predators for their size. They sometimes settle near divers in the hope that startled fish might pass within ambush range. Bougainville Reef.

Below left: Freckled Hawkfish (*Paracirrhites forsteri*). Bougainville Reef.

Below right: More than one diver has been startled by a white toothy grin set back in the dark recess of a cave. Fortunately not a dead sailor, but markings across the mantle of the Giant Thorny Oyster (*Spondylus varius*). Bougainville Reef.

Holmes Reefs

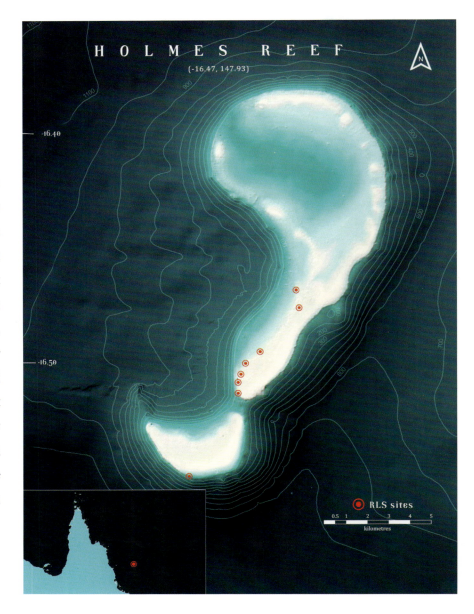

HOLMES REEF
(-16.47, 147.93)

RLS sites

0.5 1   2   3   4   5
kilometres

Three shallow reefs clustered in close proximity form the Holmes Reef system. They are located 250km east of Port Douglas, and rise from 600m water depth. Holmes Reef East encloses a large 10km-wide lagoon with water depths to 40m; however, underwater visibility is generally lower than at the other two reefs, so it is dived less frequently. Holmes Reefs West and South are separated by a narrow 500m-wide channel that drops to 80m depth. Sharks, tuna and other large pelagic fishes are seen on most dives in the channel and around the reef edges.

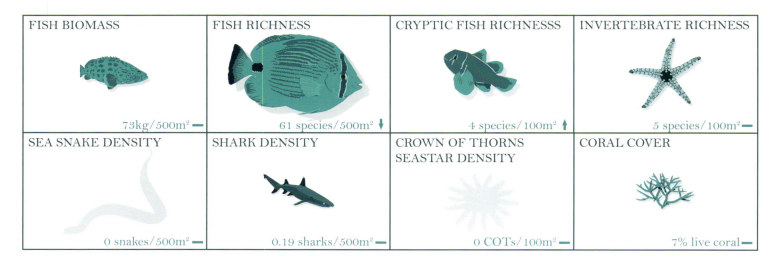

FISH BIOMASS	FISH RICHNESS	CRYPTIC FISH RICHNESSS	INVERTEBRATE RICHNESS
73kg/500m² —	61 species/500m² ↓	4 species/100m² ↑	5 species/100m² —

SEA SNAKE DENSITY	SHARK DENSITY	CROWN OF THORNS SEASTAR DENSITY	CORAL COVER
0 snakes/500m² —	0.19 sharks/500m² —	0 COTs/100m² —	7% live coral —

Several small hawkfish species live as ambush predators waiting for passing prey amongst coral heads. Holmes Reef.

Large flocks of basslets – a group of small-bodied plankton-feeding groupers – add colour to vertical walls. A single male is generally present in each flock, herding juveniles and females. Male Orange Basslet (*Pseudanthias squamipinnis*). Holmes Reef.

Sabre Squirrelfish (*Sargocentron spiniferum*) suck in water with passing fish prey by rapidly opening their large mouth cavity. Their common and scientific names both refer to the large spine projecting back from the head. Holmes Reef.

ALL PHOTOS IAN SHAW

Flinders Reefs

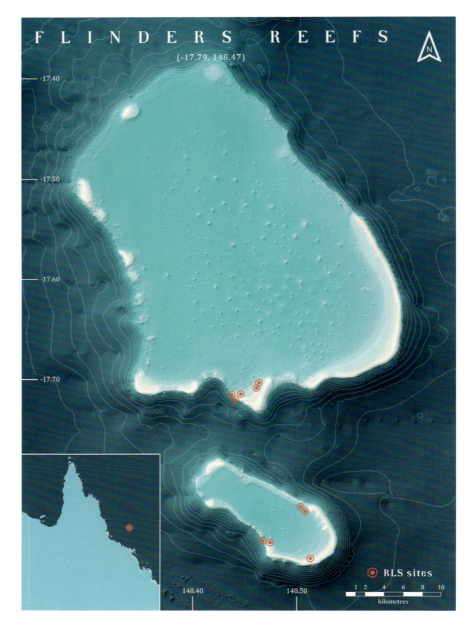

The Flinders Reef system includes remnant atolls, pinnacles and shoals located 250km east of Innisfail. North Flinders Reef – the large central reef in the system – consists of a narrow ribbon of shallow reefs set in a horseshoe pattern. These ribbon reefs protect a lagoon about 25km across that is open to the north and filled with around 50 large bommies. A cay composed of shifting sands (Main Cay), with a large automatic weather station, is present at the southern end of the reef, but waters can fully submerge this cay on extreme high tides. A 6km-wide, 600m-deep channel separates North Flinders Reef from South Flinders Reef.

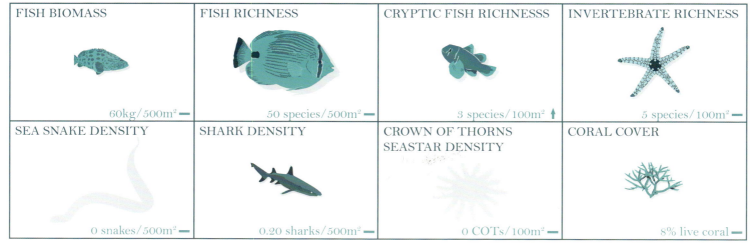

FISH BIOMASS	FISH RICHNESS	CRYPTIC FISH RICHNESSS	INVERTEBRATE RICHNESS
60kg/500m² —	50 species/500m² —	3 species/100m² ↑	5 species/100m² —

SEA SNAKE DENSITY	SHARK DENSITY	CROWN OF THORNS SEASTAR DENSITY	CORAL COVER
0 snakes/500m² —	0.20 sharks/500m² —	0 COTs/100m² —	8% live coral —

Main Cay, Flinders Reefs.

Above left: With patterning that closely matches the arrangement of *Pocillopora* polyps on which it rests, a Leopard Blenny (*Exallias brevis*) scans for small passing prey. Flinders Reefs.

Above right: Bluespotted Coral Trout (*Plectropomus leopardus*) is one of a number of large predatory fish species that stalk Coral Sea reefs. Flinders Reefs.

Below left: Reticulated Fangblenny (*Meiacanthus reticulatus*) approach fishes, nipping small pieces of their skin for food. The species is poorly known because of its highly restricted Coral Sea distribution from Flinders Reef to Osprey Reef. This fangblenny may have poison glands attached to the teeth, as in some close relatives, but this remains unknown. Flinders Reefs.

Below right: Bicoloured Dottyback (*Pictichromis coralensis*), a common species that adds colour to cracks on shaded drop-offs. Flinders Reefs.

ANDREW GREEN

ANDREW GREEN

Above left: Filter-feeding ascidian, *Polycarpa clavata*. Flinders Reefs.

Above right: Male Sailfin Queen (*Pseudanthias pascalus*). Flinders Reefs.

Below left: Deep walls on Coral Sea reefs are often undercut, with gorgonians and the ever-present Swallowtail Basslet (*Serranocirrhitus latus*). Flinders Reefs.

Below right: Tangaroa Goby (*Ctenogobiops tangaroai*) beside burrow. Flinders Reefs lagoon.

Herald Cays

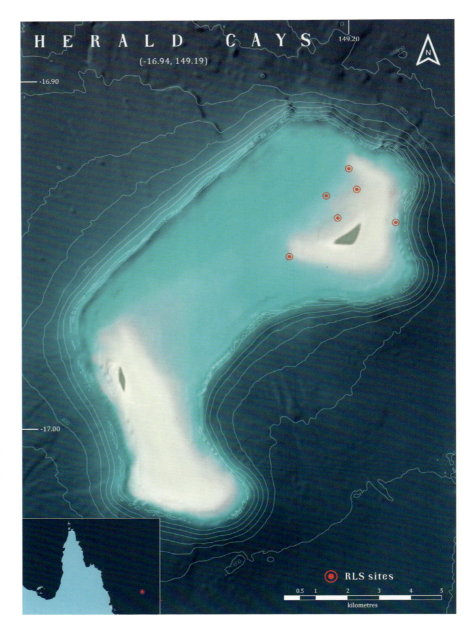

T wo islets (North East Cay and South West Cay) form this island group, which is located 350km east of Cairns. The reef system sits on a platform rising from 600m water depth. Strong currents flow between the islets through a shallow 7km-wide passage. In addition to steep drop-offs, divers are attracted to the Herald Cays because of the complex structure provided by sheltered bommies and caves on the western reef fringes. Both islets are densely vegetated and have birdlife that is exceptionally abundant.

FISH BIOMASS	FISH RICHNESS	CRYPTIC FISH RICHNESSS	INVERTEBRATE RICHNESS
69kg/500m² —	53 species/500m² ↓	3 species/100m² ↑	5 species/100m² —
SEA SNAKE DENSITY	SHARK DENSITY	CROWN OF THORNS SEASTAR DENSITY	CORAL COVER
0 snakes/500m² —	0.68 sharks/500m² —	0.03 COTs/100m² —	9% live coral —

Above: North East Herald Cay.

Below: An intertidal platform of beach rock protects the wave-exposed coast. North East Herald Cay.

Above left: Within the Coral Sea, islets with the most complex vegetation are confined to the central region. Small trees are present as well as shrubs and grasses. These attract a much greater range of birds than grassed islets, which in turn provide a greater range of nesting habitats than sand cays. North East Herald Cay includes the densest scrub thicket in the Coral Sea, with small *Pisonia grandis* trees reaching 4m in height. South West Herald Cay is also vegetated. Large populations of many bird species nest on both islands. North East Herald Cay.

Above right: Masked Boobies (*Sula dactylatra*) nesting under the coastal shrub *Heliotropium arboreum*. North East Herald Cay.

Below left: The Red-footed Booby (*Sula sula*) is unique amongst boobies in possessing grasping feet, enabling it to hold onto branches and nest in shrubs. North East Herald Cay.

Below right: Red-footed (*Sula sula*) and Brown Boobies (*Sula leucogaster*) hover over North East Herald Cay, with a squadron of frigatebirds above.

Above left: Bluespotted Coral Trout (*Plectropomus laevis*), North East Herald Cay.

Above right: Blackfin Pigfish (*Bodianus loxozonus*). North East Herald Cay.

Below left: Pair of Dwarf Hawkfish (*Cirrhitichthys falco*). North East Herald Cay.

Below right: Herbivorous blennies are common amongst coral rubble, retreating into cracks on the approach of divers and large fishes (Triplespot Blenny, *Crossosalarias macrospilus*). North East Herald Cay.

Fishing has been excluded from the Herald Cays for over 30 years, resulting in sizeable populations of large fish species that are heavily fished elsewhere. Grey Reef Shark (*Carcharhinus amblyhrhynchos*).

Lesser Queenfish (*Scomberoides lysan*) are most commonly observed around bommies within lagoons, rather than off outer reef walls where other trevally species tend to reside.

Willis Islets

The Willis Islets lie on the rim of a submerged reef system that extends over 45km north to south, at a distance of 460km from Port Douglas, and 40km from the Magdelaine Cays to the south-east. Three cays – South Islet, Mid Islet and North Cay – sit on the north-east margin of this reef system. South Islet is more commonly known as Willis Island, the only land in the Coral Sea Marine Park with human residents. Australian Bureau of Meteorology staff are stationed on the island for six-month periods running a weather station, which includes buildings, fuel tanks, and a prominent round radar dome. The station commenced operation in 1921 to provide early warning of cyclone formation in the Coral Sea following catastrophic storms that devastated the towns of Innisfail and Mackay in 1918. Willis Island is now classed as an external territory under Australian law, so has an unusual legal arrangement that requires customs and immigration clearance for visiting vessels.

The shallow Willis Island reef has a wave exposed eastern margin. The seabed west and south of the island is sandy, slopes gradually to deeper water, and accommodates a minefield of scattered bommies in all shapes and sizes.

Whitmae's Sea Cucumber (*Holothuria whitmaei*) is recognised globally as threatened (IUCN Endangered category) because of high commercial value and overharvesting, but is still moderately common on remote Coral Sea reefs. Willis Island.

119

Above: Oceanic wanderers that drift onto coral reefs often show scars resulting from attack by reef creatures. Giant Mauve Jelly (*Thysanostoma thysanura*). Willis Island.

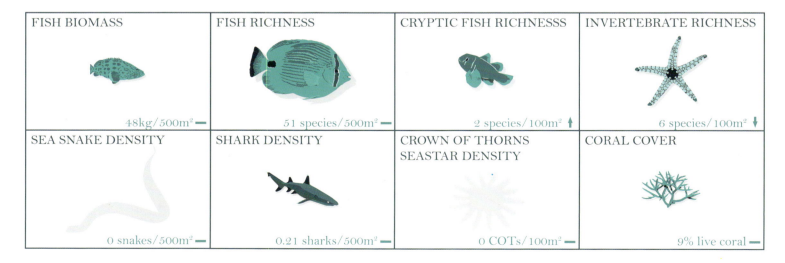

FISH BIOMASS	FISH RICHNESS	CRYPTIC FISH RICHNESSS	INVERTEBRATE RICHNESS
48kg/500m² —	51 species/500m² —	2 species/100m² ↑	6 species/100m² ↓
SEA SNAKE DENSITY	SHARK DENSITY	CROWN OF THORNS SEASTAR DENSITY	CORAL COVER
0 snakes/500m² —	0.21 sharks/500m² —	0 COTs/100m² —	9% live coral —

Blackwater diving in the Coral Sea

A largely unknown world opens up to experienced divers through 'blackwater' diving. This involves submerging offshore with torchlight when the moon is below the horizon, the sea bottom thousands of metres below. After initial disorientation – a weighted line the only point of reference – a procession of weird sea creatures come into view. Most live in the twilight zone below 200m depth during daylight but move up closer to the surface at night following their small planktonic prey. The daily cycle of plankton moving up and down between near surface waters and great depth is the largest migration on Earth, when total animal weight and distance moved are considered. Animals in all shapes and sizes, with spines, flanges, tentacles and nets, drift past the blackwater diver. Odd background whooshing and glugging sounds add further to the other-worldly nature of this domain.

Below: Ctenophore (*Ocyropsis* species), offshore Marion Reef.

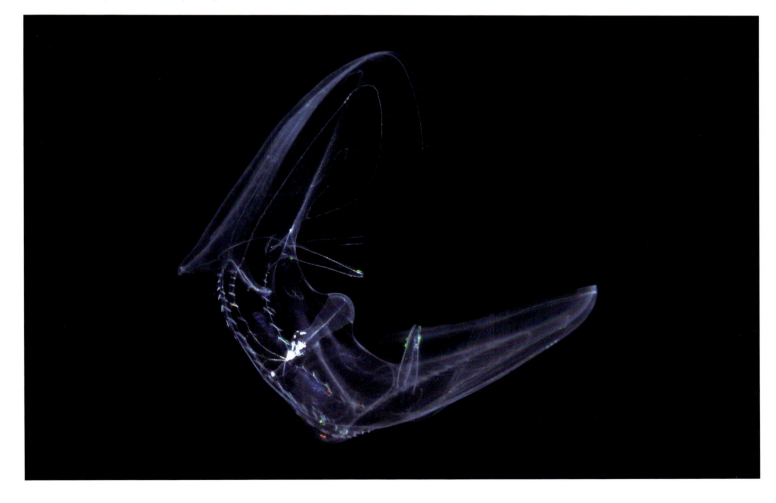

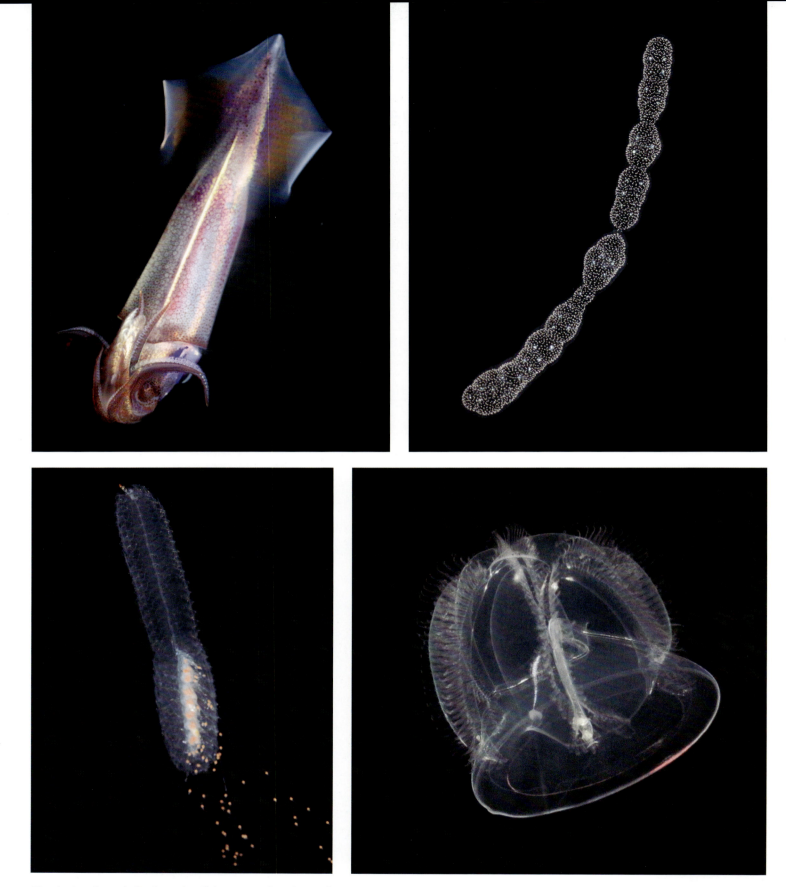

Clockwise from left: Squid, offshore Frederick Reefs; Radiolarian (*Collozoum* species), offshore Marion Reef;
Siphonophore (*Forskalia* species), offshore Frederick Reefs; Larval hemichordate, offshore Frederick Reefs.

Above: Ctenophore, offshore Marion Reef.

Below: Salp (*Pegea confoederata*), offshore Marion Reef.

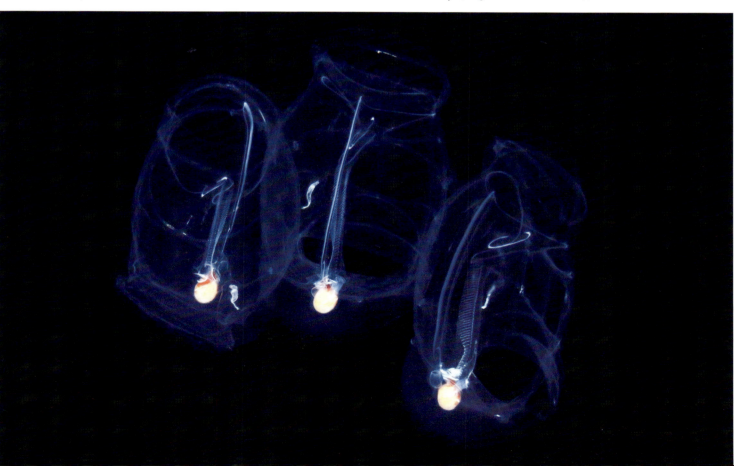

Above: Singleline Gemfish (*Promethichthys prometheus*), offshore Marion Reef.

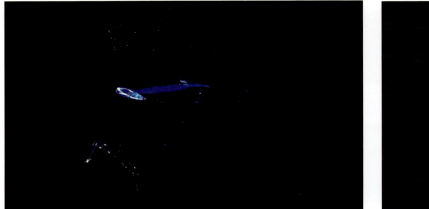

Left: Tadpole-shaped appendicularian (*Oikopleura fusiformis*) in the centre of mucous feeding net, offshore Marion Reef.

Right: Larval lobster, offshore Marion Reef.

Above: Pteropod mollusc, offshore Marion Reef.

Above: Hydromedusa, offshore Marion Reef.

Magdelaine Cays

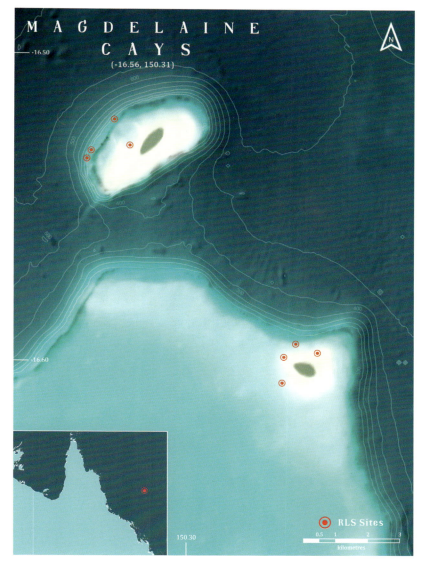

A large undersea plateau exists 300km offshore in the central Coral Sea, formed from a fragment that broke away from the Australian continental shelf. This 'Queensland Plateau' extends over 100km across, mostly with depths shallower than 200m. Numerous shallow reefs, cays and islets are scattered across the plateau, including the two Magdelaine Cays positioned 8km apart in the north-east corner. Magdelaine Cay North is devoid of vegetation, and has few seabirds. By contrast Magdelaine Cay South is heavily vegetated with shrubs as well as grasses. Frigatebirds, boobies and terns are all abundant. Coral cover has declined on both cays since the 2016 heatwave; regardless, large fishes remain common in this no-fishing location.

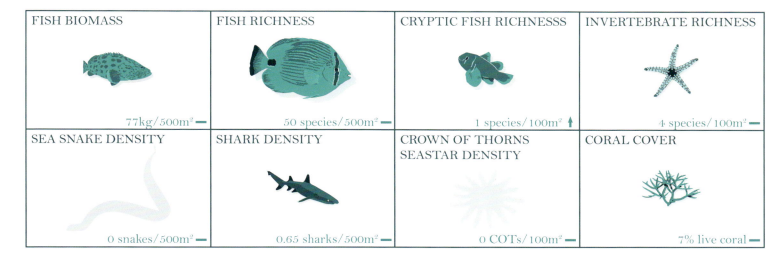

FISH BIOMASS	FISH RICHNESS	CRYPTIC FISH RICHNESSS	INVERTEBRATE RICHNESS
77kg/500m² —	50 species/500m² —	1 species/100m² ↑	4 species/100m² —
SEA SNAKE DENSITY	SHARK DENSITY	CROWN OF THORNS SEASTAR DENSITY	CORAL COVER
0 snakes/500m² —	0.65 sharks/500m² —	0 COTs/100m² —	7% live coral —

'Magdelaine Cay hosts a turtle graveyard. This was both fascinating and upsetting to visit. It looks like an idyllic nesting ground from the ocean, with a soft, sandy and sloping beachfront with plenty of space. When you walk up to the crest, it is an abrupt change, into a nightmarish battlefield of jagged rocks. Some turtles try to tackle these grounds and get trapped, unfortunately perishing.'

KIRSTY WHITMAN, RLS BLOG.

Above left: Beach rock provides feeding grounds for high densities of Natal Sally Lightfoot crabs (*Grapsus tenuicrustatus*), which forage on fine green algal turf as the tide progresses across the shore. Magdelaine Cay North.

Above right: Magdelaine Cay North is protected along its wave-exposed eastern shore by ridged beach rock. The western shore is soft sand.

Below: The exposed reef crest at Magdelaine Cay drops steeply into water over 500m deep. Seawater surges through narrow channels with few live corals present. Magdelaine Cay North.

Coringa Islets

The Coringa Islets sit on a string of reefs positioned along the north-western margin of the undersea Queensland Plateau, at 480km distance from Cairns in the west. The reefs run in a south-westerly direction from the largest islet, Chilcott, then a gap of 10km to South West Islet, then as submerged reefs. Both islets are vegetated. They provide home to dense colonies of seabirds, particularly South West Island which reaches an elevation of 14m and supports many Great Frigatebirds (*Fregata minor*) nesting in shrubs.

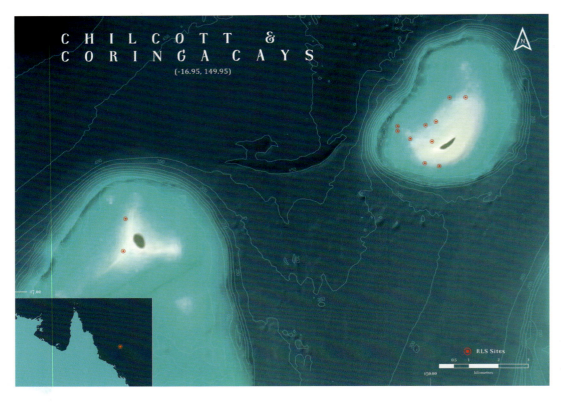

CHILCOTT & CORINGA CAYS
(-16.95, 149.95)

RLS Sites

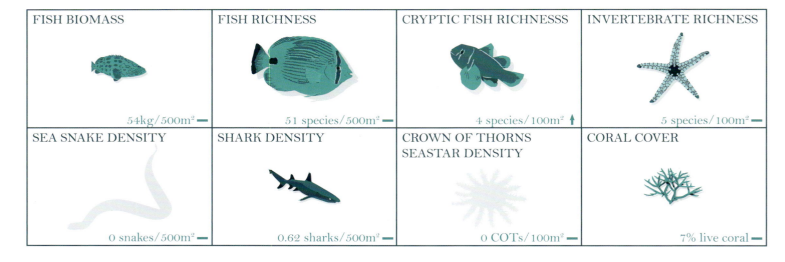

FISH BIOMASS	FISH RICHNESS	CRYPTIC FISH RICHNESSS	INVERTEBRATE RICHNESS
54kg/500m² —	51 species/500m² —	4 species/100m² ↑	5 species/100m² —
SEA SNAKE DENSITY	SHARK DENSITY	CROWN OF THORNS SEASTAR DENSITY	CORAL COVER
0 snakes/500m² —	0.62 sharks/500m² —	0 COTs/100m² —	7% live coral —

The Coringa Islets have been fully protected from fishing for over 30 years, and sharks remain common. ANDREW GREEN
Grey Reef Shark (*Carcharhinus amblyrhynchos*), Chilcott Islet.

Clown Triggerfish (*Balistoides conspicillum*), Chilcott Islet.   Whitemouth Moray (*Gymnothorax meleagris*), Chilcott Islet.

The Spiny Puller
(*Acanthochromis polyacanthus*)
is unusual amongst marine
fishes in its lengthy parental
obligations. Adults not only
look after eggs, but also
protect the dark, yellow-
striped hatchlings for many
weeks. This in turn results in
very low dispersal capacity,
little genetic exchange
between oceanic islands,
and distinctive colour forms
associated with particular
reefs. Dark colour form (top),
Diamond Island; white-tailed
colour form (middle), Ashmore
Reef; light colour form
(bottom), Chilcott Reef.

ANDREW GREEN

Above: Slender-beaked Longtom are almost impossible to see against ripples on the sea surface due to counter-shading – silver coloration underneath and darker upper body. Chilcott Reef.

Below: Most large corals conceal colourful crabs and shrimps hidden in cracks. Rust-spotted Coral Crab (*Trapezia rufopunctata*), Chilcott Reef.

Above: The South Seas Demoiselle (*Chrysiptera taupou*) is an oceanic island specialist, occurring from the Coral Sea eastward to Samoa, but absent on the nearby Great Barrier Reef. Chilcott Reef.

Below: The brilliant coloration of the Lavender Dottyback (*Cypho purpurascens*) is less striking underwater due to absorption of red light by seawater. Chilcott Reef.

ALL PICS ANDREW GREEN

Abington Reef

Abington Reef is a small isolated reef located 270km east from Tully, and over 100km from the nearest reefs – the Coringa group to the east and Flinders Reef to the west. The reef arises from the south-western corner of the submerged Queensland Plateau. It is volcano-shaped, only 3km across at the top, and with vertical walls that drop 500m on all sides. The area is poorly charted and lacks a cay for shelter, so is rarely visited by boats.

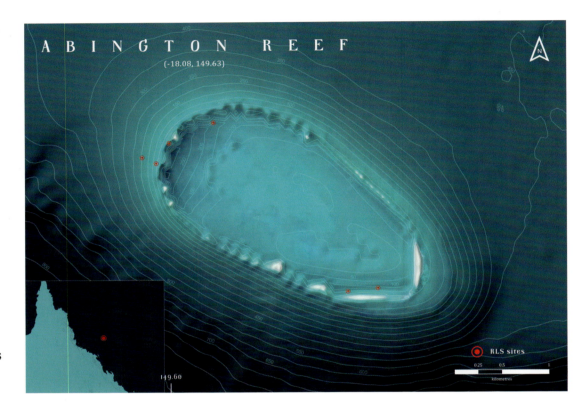

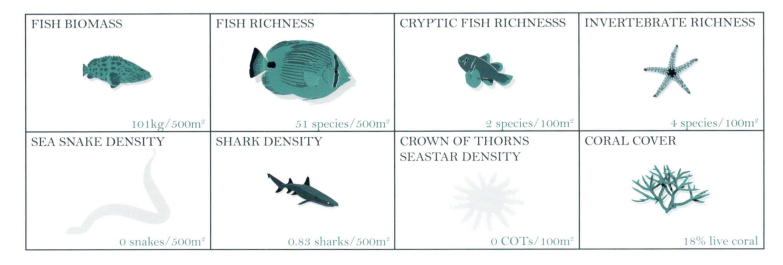

FISH BIOMASS	FISH RICHNESS	CRYPTIC FISH RICHNESSS	INVERTEBRATE RICHNESS
101kg/500m²	51 species/500m²	2 species/100m²	4 species/100m²

SEA SNAKE DENSITY	SHARK DENSITY	CROWN OF THORNS SEASTAR DENSITY	CORAL COVER
0 snakes/500m²	0.83 sharks/500m²	0 COTs/100m²	18% live coral

Barracuda (*Sphyraena* species) and other schooling pelagic fishes congregate along the reef margin. Abington Reef.

Swallowtail Basslet (*Serranocirrhitus latus*) provide colourful distractions for divers swimming along vertical walls. Abington Reef.

Diamond Islets

Four Diamond Islets (imaginatively named West, Central, East and South) are positioned around the eastern margin of a large 100km-wide atoll, 500km to the east of Innisfail. The ribbon reefs that form the outer barrier of the atoll are mostly submerged. The central lagoon basin is generally about 50m deep. As is also the case for most other cays sitting on the Queensland Plateau, the islets are vegetated and thus offer an important refuge for nesting birds.

Most visiting boats anchor at East Diamond Islet due to wonderful protection from prevailing south-easterly winds and swell behind the 8m-high islet, within a large indentation in the reef. Bommies scattered across the anchorage, some rising to near the surface from 10–40m depth, host a great variety of marine life.

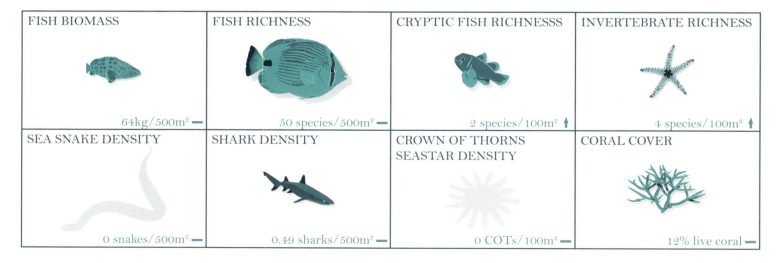

FISH BIOMASS	FISH RICHNESS	CRYPTIC FISH RICHNESSS	INVERTEBRATE RICHNESS
64kg/500m² —	50 species/500m² —	2 species/100m² ↑	4 species/100m² ↑
SEA SNAKE DENSITY	SHARK DENSITY	CROWN OF THORNS SEASTAR DENSITY	CORAL COVER
0 snakes/500m² —	0.49 sharks/500m² —	0 COTs/100m² —	12% live coral —

'The architecture of the reef out in the Coral Sea was absolutely stunning; with massive bommies, gullies and swim-throughs. The coral cover was quite low and the reef dominated by algae, however the water was warm with incredible 40m visibility. On one dive I saw six inquisitive Grey Reef Sharks (Carcharhinus amblyrhynchos), ranging in size from 1m to 2m!'

KIRSTY WHITMAN, RLS BLOG. EAST DIAMOND ISLAND.

East Diamond Island.

The central Coral Sea lies in a cyclone belt. Severe cyclones form in the region and pass each reef regularly (less than every 20 years on average). Recent heatwaves also have negatively affected coral cover. A consequence of these repeated disturbances is that the reef surface has little coral complexity. Branched and table corals need to be fast-growing to survive, and are generally rare. Stinging hydrocoral (*Millepora* species) on the reef slope. East Diamond Island.

Crown-of-thorns sea stars (*Acanthaster* species) are rare across the Coral Sea Marine Park. They generally cause less damage than heatwaves and cyclones on isolated oceanic reefs. East Diamond Islet.

Above: Shoreline, Central Diamond Islet.

Below left: *Argusia* shrubland with nesting Brown Noddies (*Anous stolidus*), East Diamond Islet.

Below right: Nesting Red-footed Boobies (*Sula sula*), East Diamond Islet.

Above: Whitetip Reef Shark (*Triaenodon obesus*), Diamond Islet.

Left: As in all Coral Sea reefs protected for fishing for several decades, large fishes are common. Spangled Emperor (*Lethrinus nebulosus*). East Diamond Islet.

Right: Blackfin Barracuda (*Sphyraena qenie*), East Diamond Islet.

'Leaving the Diamond Islets at night with the wind behind us made for a beautiful and serene sail. All the stars in the Milky Way were out and shining. The bioluminescence, stoked by the movement of Eviota, *was lighting up a beautiful green-bluish colour, sparkling out into the ripples.*'

KIRSTY WHITMAN, RLS BLOG. SOUTH DIAMOND ISLAND.

Lihou Reef

Lihou Reef is the largest atoll in Australian waters, and one of the 12 largest atolls worldwide. Its lagoon extends 100km from the south-west to north-east, covering an area of 2,529km². The outer edge drops near-vertically to over 600m depth. A string of 24 cays that together cover an area of only 1km² sit around the rim, separated by current-swept passages connecting the lagoon to open ocean. Some cays consist only of sand while five have a covering of prickly grass. The central

basin of the lagoon reaches 50m depth and is studded with bommies, many of which rise to near the sea surface. Numerous ships have run aground on the encircling ribbon reefs and are now submerged wrecks.

Due to the great distance to the mainland (600km), Lihou Reef is rarely visited and largely unexplored. A highly diverse range of habitats await divers, from spectacular bommies with good coral cover in the central lagoon to vertical walls that drop into the abyss along the outer margins. All marine species have been protected from fishing at Lihou Reef since 1982.

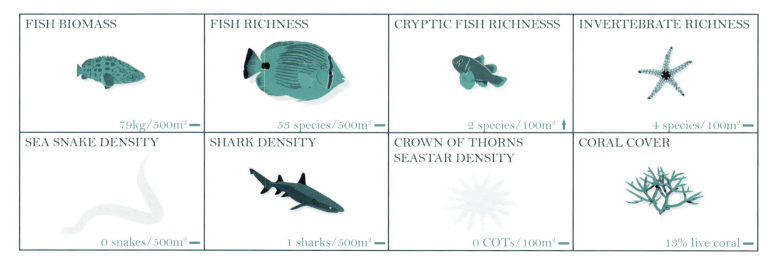

FISH BIOMASS	FISH RICHNESS	CRYPTIC FISH RICHNESSS	INVERTEBRATE RICHNESS
79kg/500m² —	53 species/500m² —	2 species/100m² ↑	4 species/100m² —
SEA SNAKE DENSITY	SHARK DENSITY	CROWN OF THORNS SEASTAR DENSITY	CORAL COVER
0 snakes/500m² —	1 sharks/500m² —	0 COTs/100m² —	13% live coral —

Observatory Cay, Lihou Reef.

Due to positioning of Lihou Reef in the cyclone belt, corals are relatively sparse, with much of the seabed covered with red calcareous encrusting algae and creeping algal forms. *Caulerpa* species, Lihou Reef.

Turtle Islet, Lihou Reef.

Left: Indian Sea Star (*Fromia indica*), Lihou Reef.

Below: Sea stars and other invertebrates that feed on algae and colonial invertebrates attached to the reef floor have few places to hide when corals are scarce. Watson's Sea Star (*Gomophia watsoni*). Lihou Reef.

Above: Bennett's Feather Star (*Anneissia bennetti*), Lihou Reef.

Below: Bicolor Fangblennies (*Plagiotremus laudandus*) are among few fishes that attack other fishes larger than themselves, nipping small pieces of flesh. Lihou Reef.

Marion Reef

Marion Reef sits isolated above an extension of the Australian continental slope, 380km east from Airlie Beach. It is an elliptical atoll rising from 400m depth, with a submerged western margin and a narrow emergent ribbon reef along its eastern margin that stops wave action. Numerous bommies rising to near sea level are scattered within the white sand lagoon. Three cays sit atop the reef – Carola in the east, Paget in the south-east, and Brodie in the south. None of the cays are vegetated, rather they consist of sands that shift as cyclones pass. Fishing is banned within the northern half of the lagoon.

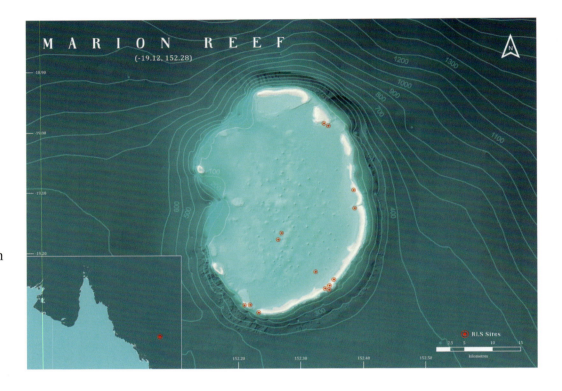

MARION REEF
(-19.12, 152.28)

● RLS Sites

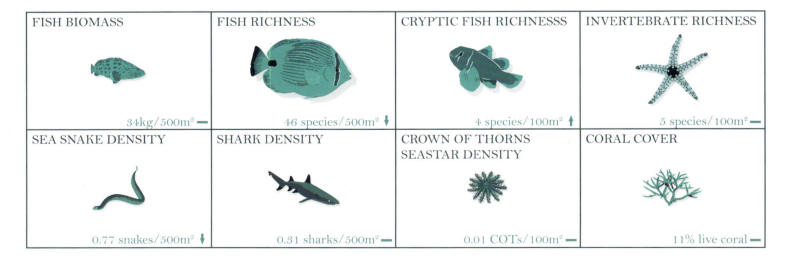

FISH BIOMASS	FISH RICHNESS	CRYPTIC FISH RICHNESSS	INVERTEBRATE RICHNESS
34kg/500m² —	46 species/500m² ↓	4 species/100m² ↑	5 species/100m² —
SEA SNAKE DENSITY	SHARK DENSITY	CROWN OF THORNS SEASTAR DENSITY	CORAL COVER
0.77 snakes/500m² ↓	0.31 sharks/500m² —	0.01 COTs/100m² —	11% live coral —

Above left: Brodie Cay, Marion Reef.   Above right: Paget Cay.

Below: Brown stain streaks across the reef surface in the foreground indicate iron pollution from a wrecked ship, in this case the *Lady Nathalia*, which ran aground beside Paget Cay in 1968. Most tropical oceanic waters have insufficient dissolved iron to support productive growth of plants. Addition of iron to these waters can cause an algal bloom, which in turn adversely affects corals through competition for space, and reef fishes because algae are not palatable. Iron wreckage has been removed from some Pacific reefs at considerable cost to eliminate this source of pollution[35].

Anemonefishes. *Amphiprion melanopus*, above; *Amphiprion perideraion*, below. Marion Reef.

Above left: Purple Brittle Star (*Ophiothrix purpurea*) hiding amongst soft coral. Marion Reef.

Above right: Stony Coral Ghostgoby (*Pleurosicya micheli*) resting on the mantle of a Fluted Giant Clam (*Tridacna squamosa*). Marion Reef.

Below left: A colourful flatworm (*Pseudoceros dimidiatus*) undulates over the Marion Reef seabed.

Below right: No sea snakes have been observed at other central Coral Sea reefs; however, several species are abundant in the Marion Reef lagoon. Horned Sea Snake (*Acalyptophis peronii*).

Shallow sands in Coral Sea lagoons conceal highly diverse communities of molluscs and other sea creatures. Huge densities of the Banded Creeper (*Rhinoclavis fasciata*) rise up onto the sediment surface at night in Marion Reef lagoon.

Predatory sediment-dwelling gastropods move about on the surface hunting other molluscs, worms and crabs. Red-mouthed Olive (*Oliva miniacea*), Marion Reef.

Harp snails are amongst the most active predators on sand flats, stalking then engulfing prey with their large foot at night. The Queensland Harp (*Harpa queenslandica*) is only known from reef lagoons in the central Coral Sea. Marion Reef.

Mellish Reef

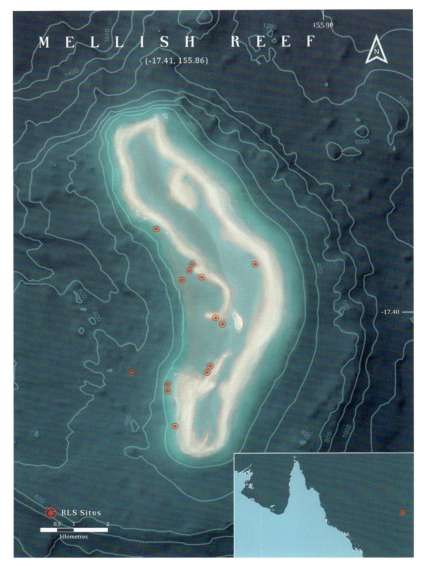

MELLISH REEF
(-17.41, 155.86)

RLS Sites
0.5 1 2
kilometres

Mellish Reef is located in the far east of the Coral Sea Marine Park at the geographical centre of the Coral Sea, almost equidistant (about 700km) from the Australian continent, New Guinea, New Caledonia and the Solomon Islands. This reef sits at the apex of a deep northward-trending volcanic ridge that rises to the surface from the 2,000m-deep bathyal plain. Very few ships visit this extremely remote location, possibly as few as one per year. Most of the shallow reefs have never been dived.

Two cays emerge above sea level. The largest (Herald Beacon Islet) lies in the centre of the reef, its surface carpeted by short grass with many nesting seabirds. A second cay positioned at the northern end of the lagoon is unstable, shifting with the movement of sands. Most of the narrow lagoon is very shallow (less than 3m). Depths exceeding 5m (maximum 13m) are only reached within the northern channel.

FISH BIOMASS	FISH RICHNESS	CRYPTIC FISH RICHNESSS	INVERTEBRATE RICHNESS
71kg/500m²	46 species/500m²	1 species/100m²	4 species/100m²

SEA SNAKE DENSITY	SHARK DENSITY	CROWN OF THORNS SEASTAR DENSITY	CORAL COVER
0 snakes/500m²	0.57 sharks/500m²	0 COTs/100m²	13% live coral

Above: Low grass patches and terns characterise the surface of Herald Beacon Islet. Mellish Reef.

Brown Booby (*Sula leucogaster*), Mellish Reef.   Coral (*Galaxea horrescens*), Mellish Reef.

Above: Bluelined Surgeonfish (*Acanthurus lineatus*) can occur in large numbers cruising across shallow reefs. They browse algae. Mellish Reef.

Below left: Giant Trevally (*Caranx ignobilis*) stalk prey within the shallow Mellish Reef lagoon.

Below right: Recreational and commercial fishing were prohibited at Mellish Reef with the introduction of a new management plan in 2018. The reef's great isolation partly protected large fishes before this time. Camouflage Grouper (*Epinephelus polyphekadion*), Mellish Reef.

Above: Rubble habitat provides shelter for a range of burrow-dwelling fishes, including the Red Firegoby (*Ptereleotris magnifica*). Mellish Reef.

Below left: Hermit crabs help maintain healthy reefs by scavenging and removing dead animal material. Swift-footed Hermit Crab (*Dardanus lagopodes*). Mellish Reef.

Below right: Lesser Queenfish (*Scomberoides lysan*), Mellish Reef.

Saumarez Reef

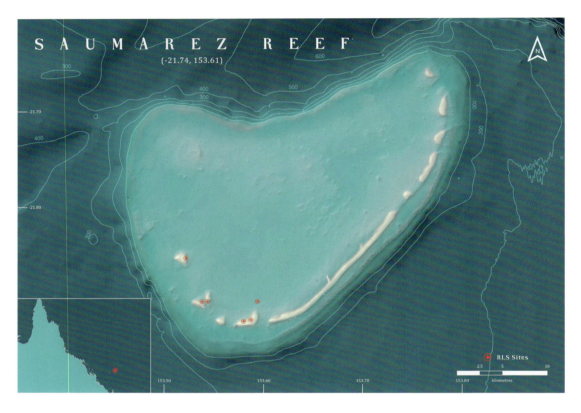

S A U M A R E Z   R E E F
(-21.74, 153.61)

RLS Sites

Like Marion Reef 300km to the north, Saumarez is a shelf-edge atoll rising from 300–400m depth on the continental slope, and with a narrow ribbon reef along the eastern margin that protects a white sand lagoon from wave action. The atoll lies 300km off the Queensland coast. A sand cay sits each end of the narrow reef, North East Cay and South West Cay. These cays provide resting places for seabirds but are too small and unstable for nesting.

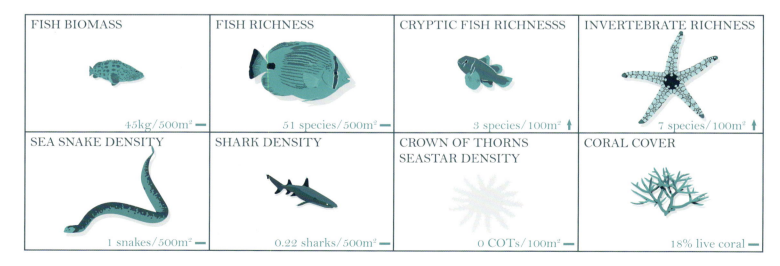

FISH BIOMASS	FISH RICHNESS	CRYPTIC FISH RICHNESSS	INVERTEBRATE RICHNESS
45kg/500m² —	51 species/500m² —	3 species/100m² ↑	7 species/100m² ↑

SEA SNAKE DENSITY	SHARK DENSITY	CROWN OF THORNS SEASTAR DENSITY	CORAL COVER
1 snakes/500m² —	0.22 sharks/500m² —	0 COTs/100m² —	18% live coral —

Above: Turtle Weed (*Chlorodesmis fastigiata*).
Saumarez Reef.

Below left: Bommies within lagoons across the Coral Sea host a diverse range of green algal species, particularly on the shaded faces of vertical walls. Many can creep rapidly across the seabed by extending root-like stolons, including species of *Caulerpa*. Fern-like rosettes of *Caulerpa filioides* blossom amongst other green algae, Saumarez Reef.

Below right: *Udotea orientalis*, a distinctive fan-shaped green alga. Saumarez Reef.

A great variety of sessile (attached) filter-feeding invertebrates occur within the Coral Sea Marine Park, generally in shaded locations with good current flow. Whereas a number of specialist biologists are working to identify the reef-building corals, most soft corals, sponges, ascidians and bryozoans remain unnamed and largely unknown. All Saumarez Reef.

Above: Soft coral (*Lobophytum* species).

Below: Soft coral (*Lobophytum* species).

Above: Soft coral (*Congomeratusclera* species).

Below: Soft coral.

Above: Sponge.

Below: Sponge.

Above: Sponge.

Below: Bryozoan.

Bicolour Parrotfish (*Cetoscarus ocellatus*) undergo an immense transformation in appearance from juvenile (below) to adult (above). Saumarez Reef.

Above: Anemones as well as corals expel symbiotic algae from their tissues and turn white ('bleach') during heatwaves. Leathery Anemone (*Heteractis crispa*) and Orangefin Anemonefish (*Amphiprion chrysopterus*), Saumarez Reef.

Below: Starry Cup Coral (*Acanthastrea echinata*), Saumarez Reef.

Above: Mottled colour patterning on the Tripletail Maori Wrasse (*Cheilinus trilobatus*) breaks up the body outline, camouflaging it amongst the random assortment of seabed organisms. Saumarez Reef.

Below left: Large hermit crabs (*Dardanus megistos*) scavenge across the lagoon seafloor at night. Saumarez Reef.

Below right: Pink-clawed Hermit Crab (*Dardanus pedunculatus*) releasing brooded juveniles from the upper shell opening. Saumarez Reef.

Left: Bigfin Reef Squid (*Sepioteuthis lessoniana*) patrol the Saumarez lagoon at night.

Right: Banded Coral Shrimp (*Stenopus hispidis*) set up cleaner stations on inshore bommies to pick parasites off visiting fishes. Saumarez Reef.

Left: Robust feather star perched on fire coral (*Stylaster* species). Saumarez Reef.

Right: The Prickly Redfish (*Thelenota ananas*), one of several sea cucumber species common in the Coral Sea Marine Park, is threatened globally because of overharvesting elsewhere. Saumarez Reef.

Frederick Reefs

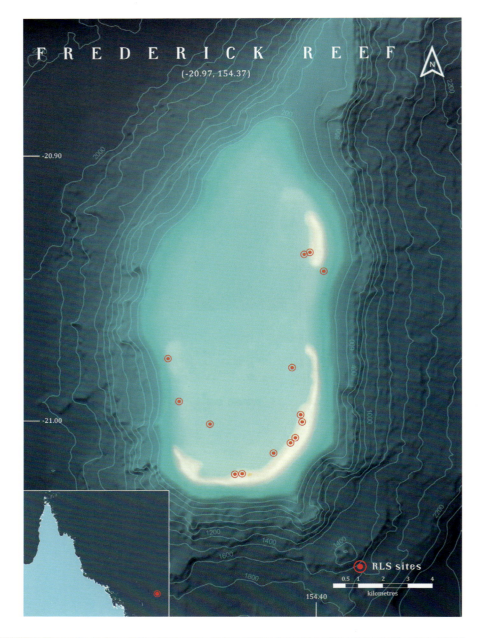

FREDERICK REEF
(-20.97, 154.37)

RLS sites

kilometres

Frederick Reef lie 500km east of Mackay, a remnant atoll isolated from all other reef systems by deep water (greater than 2,000m depth). Shallow ribbon reefs that rise to low-tide level are located along the eastern and southern atoll margins, where they block the prevailing swell from entering a 30m-deep lagoon. Two sand cays are present: Observatory Cay in the south, and a small unstable cay with an 'Eye-of-Sauron' concrete lighthouse on a circular rock plinth, set in the north-east.

FISH BIOMASS	FISH RICHNESS	CRYPTIC FISH RICHNESSS	INVERTEBRATE RICHNESS
38kg/500m² —	46 species/500m² —	2 species/100m² ↑	5 species/100m² —

SEA SNAKE DENSITY	SHARK DENSITY	CROWN OF THORNS SEASTAR DENSITY	CORAL COVER
0.75 snakes/500m² —	0.38 sharks/500m² —	0 COTs/100m² —	15% live coral —

Above: Observatory Cay, formed from coarse shell fragments and sand, rises to only 2m in height above sea level. This is, however, the only stable land surface for more than 100km in any direction, so is appreciated by resting seabirds, particularly noddies. Frederick Reefs.

Below: The Frederick Reef lagoon provides home for large populations of at least three sea snake species, including Dubois' Sea Snake (*Aipysurus duboisii*).

Left: Although open to line and spear-fishing, large predatory fishes remain relatively common at Frederick Reefs. Green Jobfish (*Aprion virescens*).

Right: Paddletail Seabream (*Gymnocranius euanus*), Frederick Reefs.

Left: The Pink Whipray (*Pateobatis fai*) excavates large pits in the sandy seabed. It can detect small electric pulses radiating from fishes, crustaceans and mollusc hidden under a shallow sand layer. Frederick Reefs.

Right: Unusually amongst squirrelfishes and soldierfishes, Slender Squirrelfish (*Neoniphon sammara*) are frequently seen by divers during daylight hours. They hover at the entrances of crevices and the edges of coral thickets. Frederick Reefs.

Above: Common Lionfish (*Pterois volitans*), Frederick Reef lagoon.

Above: Regal Angelfish (*Pygoplites diacanthus*), Frederick Reefs.

GERMAN SOLER

Above left: A curious Brown Tang (*Zebrasoma scopas*), a common species that grazes turf algae on bommies in Frederick Reef lagoon.

Above right: Yellow-stripe Goatfish (*Mulloidichthys flavolineatus*).

Kenn Reefs

KENN REEFS
(-21.20, 155.76)

RLS sites

Amongst reefs in the Coral Sea Marine Park, the Kenn Reefs are the second furthest offshore, 600km east of Mackay. Like neighbour Frederick Reefs (100km distant), Kenn Reefs sit as an isolated fragmented atoll on an extinct underwater volcano, with a base over 1km deep. The narrow reef margin exposed at low tide forms an arrowhead aimed at the most easterly point, and with shallow gaps in the reef on both flanks. A small sand cay (Observatory Cay) is positioned near the arrowhead, and a second cay (Boulder Cay) on the south-western arm.

The Kenn Reefs have superb diving, with crystalline waters, diverse marine life, large pelagic fishes, and bommies with complex structure rising up from the sandy lagoon seabed to 25m depth. The southern half of the reef remains open to fishing while the northern sector is closed.

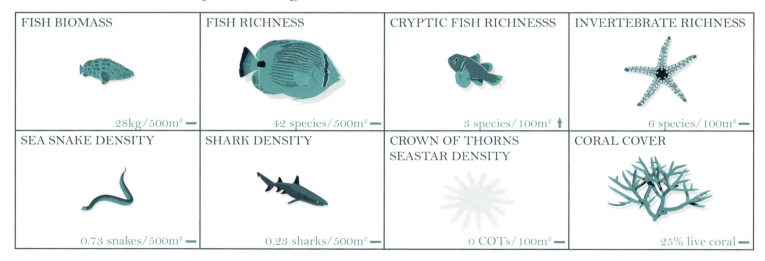

FISH BIOMASS	FISH RICHNESS	CRYPTIC FISH RICHNESSS	INVERTEBRATE RICHNESS
28kg/500m² —	42 species/500m² —	3 species/100m² ↑	6 species/100m² —
SEA SNAKE DENSITY	SHARK DENSITY	CROWN OF THORNS SEASTAR DENSITY	CORAL COVER
0.73 snakes/500m² —	0.23 sharks/500m² —	0 COTs/100m² —	25% live coral —

Above: Loggerhead Turtles (*Caretta caretta*) are regular oceanic visitors to Kenn Reefs.

Below: Many molluscs within Coral Sea lagoons hide under the white sand in daylight, later creeping along in the open at night. Vomer Conch (*Euprotomus vomer*). Kenn Reefs.

Above: Juvenile urchins (*Diadema savignyi*) emerge from cracks in the reef at night. Kenn Reefs.

Below: The Slate Pencil Sea Urchin (*Heterocentrotus mamillatus*) is more reluctant than many other urchins to leave crevices. Kenn Reefs.

Above: Ambush predators resting on reefs are found at different levels of the food chain. This Birdwire Rockcod (*Epinephelus merra*, left) would happily wolf down a Largemouth Threefin (*Ucla xenogrammus*, right), which in turn attacks small crustaceans. Kenn Reefs.

Above: Gigantic Anemone (*Stichodactyla gigantea*) are found in a variety of colours. Kenn Reefs.

Wreck Reefs

The eight small Wreck Reefs extend 30km east-west on a submerged volcanic peak in the southern Coral Sea, 500km east of Yeppoon. Four cays are present, two unvegetated sand cays in the west (West Island and Hope Cay) and two vegetated islets in the east (Porpoise Cay and Bird Islet). All have large seabird populations, with exceptional numbers of nesting birds on Bird Islet – no surprise given its name.

Underwater scenery within the Wreck Reef complex is extremely varied, arguably the most interesting for divers in the Coral Sea Marine Park. Coral cover is higher than elsewhere, and water clarity is exceptional. Although open to recreational and professional line fishing, Queensland Groper can still be seen, and sharks remain common. Porpoise Cay has great historical significance, as the site where two colonial vessels, the *Porpoise* and *Cato*, ran aground in 1803. The navigator Matthew Flinders was aboard and took command, sailing to Sydney in an open cutter to arrange rescue. A permit is needed to inspect anchors and other historical wreckage still strewn over the reef nearby.

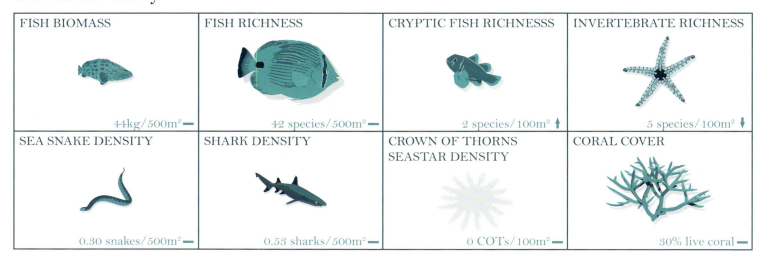

FISH BIOMASS	FISH RICHNESS	CRYPTIC FISH RICHNESSS	INVERTEBRATE RICHNESS
44kg/500m² —	42 species/500m² —	2 species/100m² ↑	5 species/100m² ↓
SEA SNAKE DENSITY	SHARK DENSITY	CROWN OF THORNS SEASTAR DENSITY	CORAL COVER
0.30 snakes/500m² —	0.53 sharks/500m² —	0 COTs/100m² —	30% live coral —

Above: Porpoise Cay has very low vegetation cover, and fewer nesting seabirds than on Bird Islet.

Below: Seabirds returning to their nests at dusk. Bird Islet.

RICK STUART-SMITH

Above: Moulting juvenile Masked Booby. Bird Islet.

'This is just below the main cyclone belt, so with great coral as well as sea snakes. No doubt about its name (Bird Islet) — the tussock-covered cay was alive with seabirds, including noddies, terns, frigatebirds, boobies, and one lost chick.'

CAITLIN KUEMPEL, RLS BLOG.

The dominant invertebrate on most Coral Sea cays is the Red Land Hermit Crab (*Coenobita perlatus*). During the day they crowd together in whatever shade is available, then move about independently looking for food as night approaches. Bird Island.

This Brown Booby (*Sula leucogaster*) was curious at seeing an intruding diver, peering down from the sea surface. Kenn Reefs.

Juveniles of the Mimic Surgeonfish (*Acanthurus pyroferus*) engage in an odd form of mimicry, with an appearance nearly indistinguishable from local angelfishes. In the Coral Sea Marine Park they mimic the Lemonpeel (*Centropyge flavissima*, below) and the Yellow (*Centropyge heraldi*) Angelfishes, whereas in the Indian Ocean, where those two angelfish species do not occur, they mimic the dissimilar Pearlscale Angelfish (*Centropyge vrolikii*). The adaptive value of pretending to be an angelfish is not known, but perhaps relates to the long defensive spine on the angelfishes' operculum, or perhaps their tendency to feed close to coral cover, which makes them difficult for predators to catch. Wreck Reef, Coral Sea Marine Park.

RICK STUART-SMITH

RICK STUART-SMITH

Yellow Angelfish (*Centropyge heraldi*), Cato Reef, Coral Sea Marine Park.

Yellowpeel Angelfish (*Centropyge flavissima*), Marion Reef, Coral Sea Marine Park.

RICK STUART-SMITH

Juvenile Mimic Angelfish (*Acanthurus pyroferus*), Ashmore Reef Marine Park.

Pearlscale Angelfish (*Centropyge vrolikii*), Ashmore Reef Marine Park.

RICK STUART-SMITH

Queensland Groper resident in gulch, Bird Islet lagoon.

Above: Queensland Groper (*Epinephelus lanceolatus*), Bird Islet lagoon.

Below: Patch reefs, Bird Islet lagoon.

Above: Bluefin Trevally (*Caranx melampygus*) patrolling shallows in Bird Islet lagoon.

Below: Small isolated bommie, Bird Islet lagoon.

Left: A juvenile Brown Tang (*Zebrasoma scopas*) – backbone showing – hovers almost invisible against the sandy lagoon floor at night. Its transparency indicates recent settlement from the plankton. Wreck Reefs.

Below: The Wreck Reefs are located below the cyclone belt, so coral reef structure develops over a long period without catastrophic disturbance. Bird Islet reef.

RICK STUART-SMITH

Above: Fast-growing stinging corals (*Millepora* species) reach above hard corals to trap passing plankton. Wreck Reefs.

Following spread: Complex coral reef structure includes a diversity of holes, cracks and fissures where moray eels and crevice-dwellers reside. Whitemouth Moray (*Gymnothorax meleagris*), Wreck Reefs.

Left: Vanderbilt's Damselfish (*Chromis vanderbilti*), named after George Vanderbilt – a scientific explorer and amongst the richest Americans – who led expeditions in the Pacific and first collected this species. Wreck Reefs.

Right: Whitetip Reef Shark (*Trienadon obesus*).

Below: A common nocturnal grazer on shallow reefs with moderate wave action, the Pacific Urchin (*Echinometra mathaei*), Wreck Reefs.

Spotted Sea Cucumber (*Synapta maculata*), Porpoise Reef.

After a day at Bird Islet, with four sites completed, we moved to nearby Porpoise Cay for another three sites. Matthew Flinders and others made a miraculous escape when their vessel, HMS Porpoise, was wrecked here in 1803, spending weeks stranded on the small cay before rowing in an open boat all the way to Sydney. Though there was no sign of wreckage in the lagoon while diving, we did find a huge snake-like holothurian under the transect line that was just as exciting.'

CAITLIN KUEMPEL, RLS BLOG.

Cato Reef

Cato Reef – the southernmost reef in the Coral Sea Marine Park – lies 110km south of the Wreck Reefs and 440km east of Gladstone. The elliptical reef encloses a shallow lagoon 2km across, studded with small bommies. Cato Island, a 6m-high vegetated cay, sits in the western quarter.

Although only 700m long, this is the largest island in the Coral Sea Marine Park, home to thousands of nesting Wedge-tailed Shearwaters, Sooty Terns and Brown Noddies, and hundreds of Lesser Frigatebirds and Masked Boobies. Severe cyclones rarely progress this far south, consequently coral cover is very high, and underwater scenery exceptional.

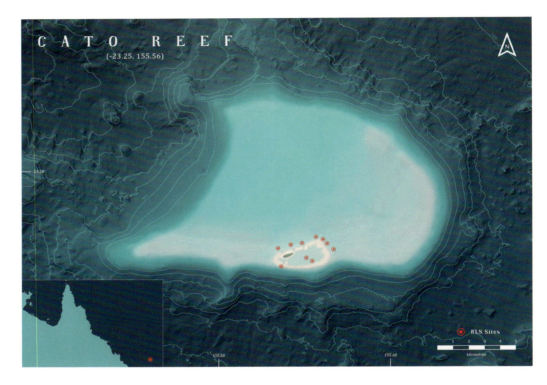

FISH BIOMASS	FISH RICHNESS	CRYPTIC FISH RICHNESSS	INVERTEBRATE RICHNESS
31kg/500m² —	43 species/500m² —	2 species/100m² ↑	5 species/100m² ↑
SEA SNAKE DENSITY	SHARK DENSITY	CROWN OF THORNS SEASTAR DENSITY	CORAL COVER
0.60 snakes/500m² —	0.03 sharks/500m² —	0 COTs/100m² —	38% live coral —

Above: A multitude of Wedge-tailed Shearwaters (*Ardenna pacifica*) returns at dusk, Cato Island.

Below left: This small hydromedusa drifting across Cato Reef is the alternate life-history stage of a hydroid attached to the seabed.

Below right: Ringeye Hawkfish (*Paracirrhites arcatus*) sitting amongst coral branches, Cato Reef.

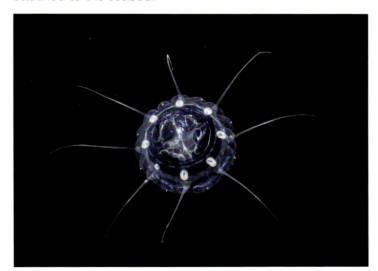

Above: Masked Booby (*Sula dactylatra*) dot the landscape, Cato Island.

Below: A characteristic feature of reef slopes is high coral cover. Branched coral (*Pocillopora eydouxi*) with small Flame Hawkfish (*Neocirrhites armatus*) hiding at centre, Cato Reef.

Above: A Johnston Damsel (*Plectroglyphidodon johnstonianus*) defends its branching coral territory from intruders much larger than itself. Cato Reef.

Below: The Princess Damsel (*Pomacentrus vaiuli*) occurs abundantly across the Coral Sea Marine Park, but is only seen occasionally on the nearby Great Barrier Reef. Cato Reef.

Nudibranchs and sapsuckers

The nudibranchs and sapsuckers are two closely related groups of shell-less molluscs with reputations as the fashion icons of the marine world due to extreme multi-coloured body patterns. The brilliant colours warn predators that tissues are laden with toxic chemicals and the animal is unpalatable. In addition to vivid design, nudibranchs can be recognised by external gills arranged along both sides of the body or in a cluster towards the back of the upper body surface. Hundreds of species live within the Coral Sea Marine Park. Close relatives, the sapsuckers, have elongate bodies and leaf-like projections on the sides. They use specialised teeth to stab and extract cellular fluids from seaweeds, mostly green algae.

'This led to amazing diving conditions, with calm clear waters and spectacular visibility (more than 40m was the norm). Some of the highlights for me included a friendly Queensland Grouper, tons of curious sea snakes, walking sharks, and three sightings of the black-and-gold sapsucking sea slug, Cyerce nigricans.'

CAITLIN KUEMPEL, RLS BLOG. FREDERICK REEFS.

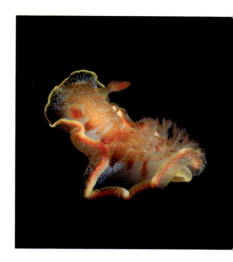

Left: Elisabeth's Nudibranch (*Chromodoris elisabethina*), Lihou Reef. Middle: Willan's Nudibranch (*Chromodoris willani*), Lihou Reef. Right: Juvenile Spanish Dancer (*Hexabranchus sanguineus*) undulating above the reef. Wreck Reefs.

Left: Bullock's Chromodorid (*Hypselodoris bullocki*), Kenn Reefs. Middle: Pustulose Nudibranch (*Phylidiella pustulosa*), Saumarez Reef. Right: Heavenly Nudibranch (*Phyllidia coelestis*), Bougainville Reef.

BOB EDGAR

Left: Neon Sapsucker (*Thuridilla neona*), Saumarez Reef. Middle: Bayer's Sapsucker (*Thuridilla bayeri*), Saumarez Reef. Right: Much-desired Flabellina (*Coryphellina exoptata*), Frederick Reefs.

South-west Marine Parks Network

A total of 14 marine parks distributed from Kangaroo Island (South Australia) to Shark Bay (central Western Australia) form the South-west Network. These parks cover 508,371km². They encompass offshore waters off the western half of the 'Great Southern Reef', the linked series of temperate rocky reefs – often kelp covered – that extend from New South Wales to Western Australia and around Tasmania[1].

Aboriginal people of south-western Australia and South Australia have been managing Sea Country that overlaps South-west Marine Parks for tens of thousands of years, including before rising sea levels created these marine environments.

South-west Marine Parks are popular destinations for fishing and boating, indirectly supporting tourism, commercial fishing, mining, and shipping activities. Most of the network overlays deep water, consequently limited opportunities exist for recreational activities other than fishing and nature-viewing tourism. Charter boats depart from a number of locations for whale watching. This activity is seasonal, primarily capitalising on the autumn migration of Humpback Whales from Antarctica to breeding grounds off north-western Australia, and, later in the year, the spring migration in return.

Amongst the most important locations for marine mammals Australia-wide are the

KEVIN SMITH

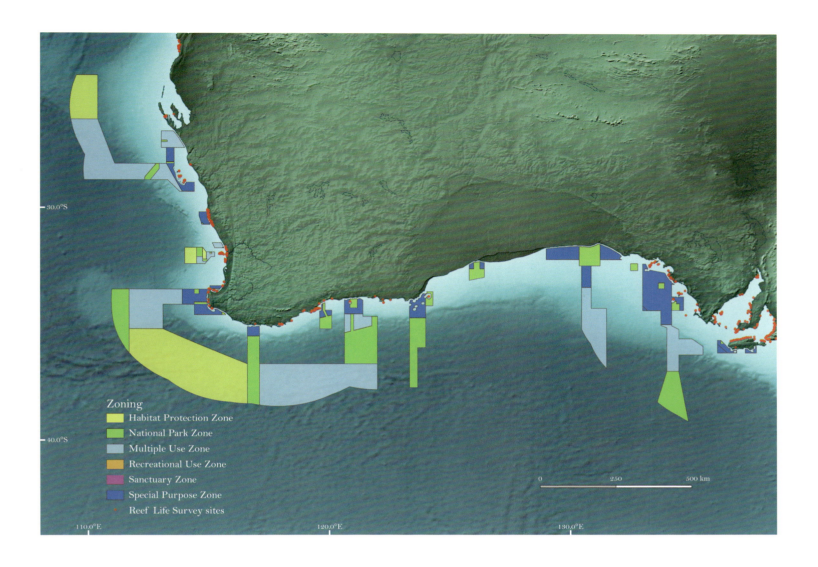

Zoning

- Habitat Protection Zone
- National Park Zone
- Multiple Use Zone
- Recreational Use Zone
- Sanctuary Zone
- Special Purpose Zone
- Reef Life Survey sites

Albany Canyons, located offshore from Fitzgerald National Park near Bremer Bay on the south coast. These shelf-edge canyons are included within the Bremer Marine Park. They funnel cool nutrient-rich waters up from the depths, greatly enhancing the productivity of planktonic organisms in nearshore waters, which in turn attracts cetaceans including Southern Right Whales and orcas, seals, and seabirds.

The largest shelf-edge canyon on the west coast, the Perth Canyon, located directly off Perth and Rottnest Island, also plays a critically important ecological role for whales and seabirds. It is protected within the Perth Canyon Marine Park.

Few shallow reefs suitable for recreational diving are located within South-west Marine Parks. The exception is the Geographe Marine Park, located 180km south of Perth. The seabed through most of this marine park is covered by beds of strap-like *Posidonia* seagrass; however, broken reef habitat with rarely seen plants and animals is also present.

Opposite: After seadragons, the Blue Devil Fish (*Paraplesiops meleagris*) is arguably the most iconic fish species inhabiting the Great Southern Reef, due to its strikingly intense coloration. Geographe Marine Park.

Geographe Marine Park

Geographe Marine Park covers most of Geographe Bay, abutting the state-managed Ngari Capes Marine Park that extends 3 nautical miles out from the shoreline[36]. The marine park lies off the towns of Bunbury and Busselton, includes 977km² total area, and extends from depths of 15–70m. Most of the marine park is open to recreational and most commercial fishing, although a small area (15km²) near Busselton is closed to fishing.

The straplike seagrass *Posidonia sinuosa* covers about 60 per cent of the seabed within Geographe Bay. Running parallel to the shore, and starting at a depth of 16m, a series of ancient hardened sand dunes outcrop in the seagrass as low narrow rocky ridges. These reefs are heavily eroded with undercuts and caves. They support seaweed and sessile invertebrate communities that are amongst the richest worldwide in terms of the number of species present in a small area.

The Noongar people have cultural responsibilities for Sea Country in the Geographe Marine Park that are passed down from elders. These include keeping their Sea Country healthy, supporting their spiritual wellbeing, and upholding and protecting their cultural responsibilities for future generations.

Zone Name	Zone Area (km²)
Marine National Park Zone (no-fishing)	15
Habitat Protection Zone (recreational and limited commercial fishing)	21
Special Purpose Zone (Mining Exclusion) (recreational and commercial fishing)	650
Multiple Use Zone (recreational and commercial fishing)	291

FISH BIOMASS	FISH RICHNESS	CRYPTIC FISH RICHNESSS	INVERTEBRATE RICHNESS
37kg/500m² —	28 species/500m² ↑	4 species/100m² —	6 species/100m² —

SEA SNAKE DENSITY	SHARK DENSITY	CROWN OF THORNS SEASTAR DENSITY	CORAL COVER
0 snakes/500m² —	0.36 sharks/500m² —	0 COTs/100m² —	3% live coral —

Normally resident in deeper water, the Yellowspotted Boarfish (*Paristiopterus gallipavo*) is regularly observed on Geographe Marine Park reefs.

'The sites extend east to west across the bay roughly in a line parallel to the coast in 16–19m. The main feature of each site is a linear reef, sometimes straight, sometimes more 'squiggly'. These reefs are characterised by fractured edges, vertical walls, small caves and undercuts, and isolated bommies. The flat terrain above is populated by low but diverse algae dotted with stony corals and sponges. Seagrass beds extend into the distance below. In some places, algae gives way to gardens of coral.'

KEVIN SMITH, RLS BLOG.

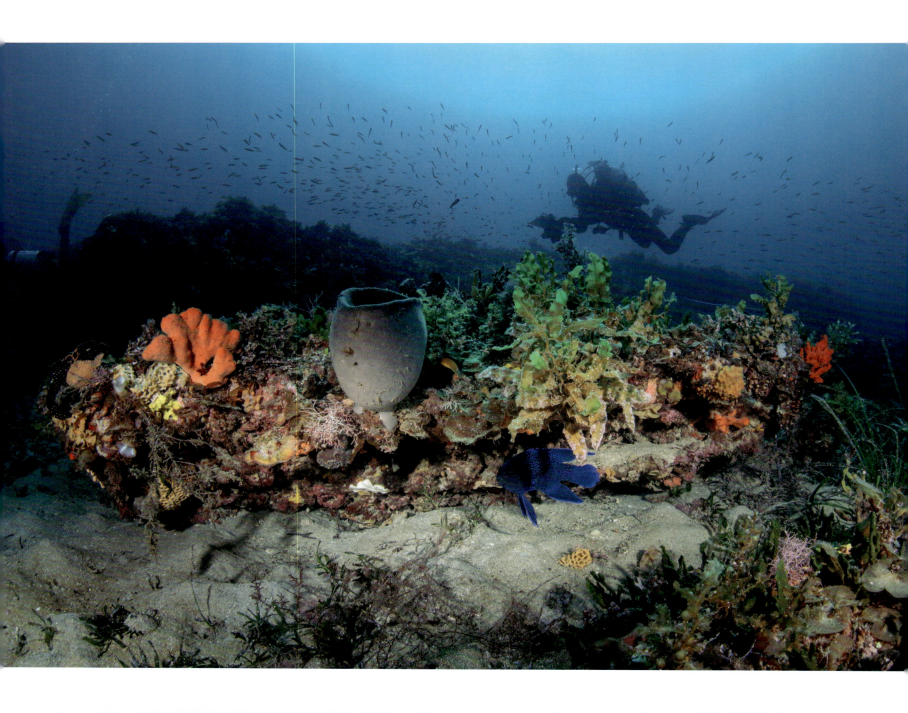

'The coral garden at Squiggly Reef is impressive, the abundance of schooling species such as Rough Bullseye (Pempheris klunzingeri) is in the thousands and Yellow-headed Hulafish (Trachinops noarlungae) in the tens of thousands. Rare and cryptic species can be sighted and iconic species are common.'

PAUL DAY, RLS BLOG.

Above: The Bluelined Leatherjacket (*Meuschenia galii*) is amongst the most curious of fishes. Geographe Marine Park.

Below: Two sea cucumbers (*Ceto cuvieri*) perch high on a massive sponge, tentacles open to catch passing plankton. Geographe Marine Park.

Above: When defending territories, Giant Cuttlefish (*Sepia apama*) aggressively approach divers. Geographe Marine Park.

Below: An unusual south-western grouper, Harlequin Fish (*Othos dentex*) generally sit immobile, perched in ambush mode. Geographe Marine Park.

Above and below: A number of tropical corals drift as larvae in the Leeuwin Current as far south as Geographe Bay Marine Park, where they form small coral patches.

Below: The Western Rock Lobster (*Panulirus cygnus*) is the most valuable fishery species in Australia. Where abundant, it also plays a large ecological role on reefs and nearby seagrass beds by controlling numbers of invertebrate prey[37]. Geographe Marine Park is at the southern edge of the rock lobster's range, so it is less often seen here than in other South-west Australian Marine Parks further north. Rottnest Island.

Below: The Western Staircase Sponge (*Caulospongia biflabellata*), a species confined to south-western Australian waters, is amongst the most eye-catching of sponges. Geographe Marine Park.

Above: Low limestone reef in Geographe Marine Park gives way along its edge to seagrass, the predominant habitat within the bay.

Below: Seagrass meadows (*Amphibolis griffithii*) with passing school of Silver Trevally (*Pseudocaranx georgianus*), Geographe Marine Park.

Temperate East Marine Parks Network

The Temperate East Marine Parks Network includes Commonwealth waters from the southern limit of the Great Barrier Reef to the southern coast of New South Wales, eastward to encompass Lord Howe Island, and includes Commonwealth waters surrounding Norfolk Island in the far east. This network spans subtropical and temperate waters, and contains extensive seamount chains[38].

The region's oceanography and marine ecology are driven by the East Australian Current, which flows southwards from the Coral Sea to waters east of Tasmania. Its strength varies seasonally and is erratic to the south, with large wobbles causing eddies and upwellings. The current flows to 500m depth and 100km width, conveying tropical and subtropical species southward to temperate waters.

Eight marine parks form the Temperate East Network, covering a total area of 383,339km². They encompass a great variety of tropical to temperate ecosystems – including seamounts, undersea ridges, abyssal basins, canyons and coral atolls – in depths ranging from the high tidal zone to 5,000m. Activities allowed in these marine parks are outlined in a management plan that commenced in 2018.

Australian Marine Parks in the Temperate East Network include parks that Traditional Owners have Sea Country connections to. Parks in this Network have a long history of human usage, including by Aboriginal people, who once lived on areas that became submerged at the end of the last glaciation. The Temperate East Network also supports commercial and

RICK STUART-SMITH

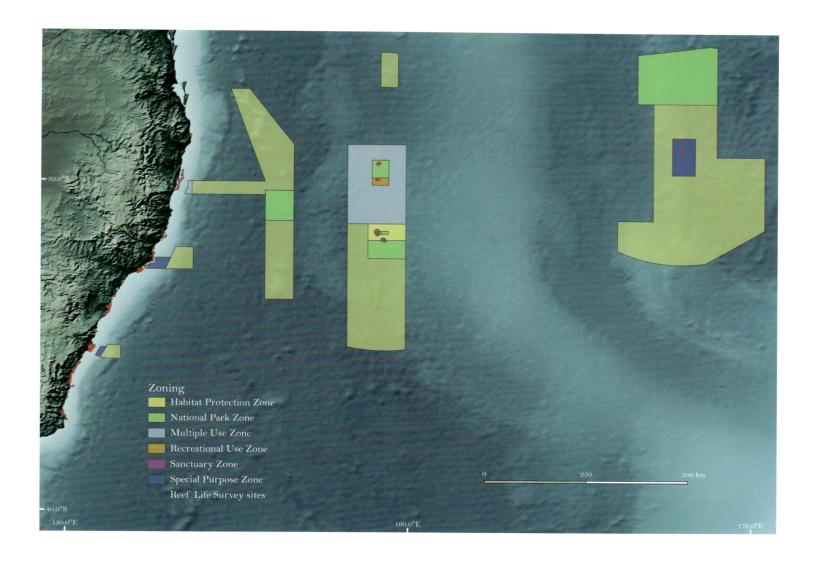

Zoning
- Habitat Protection Zone
- National Park Zone
- Multiple Use Zone
- Recreational Use Zone
- Sanctuary Zone
- Special Purpose Zone
- Reef Life Survey sites

recreational fishers, whale and seabird watchers, recreational divers, and shipping and military operations. Norfolk Island is also home to descendants of Pitcairn islanders, with Pacific Island heritage, that settled on Norfolk Island in 1856.

Shallow diveable reefs are located in four Temperate East Marine Parks – Solitary Islands, Cod Grounds, Lord Howe and Norfolk. Within the Solitary Islands Marine Park, an extraordinary subsea pinnacle (Pimpernel Rock) was protected from fishing within a 1.3km-wide sanctuary zone box in 2003. A second small no-fishing zone (2km on each side) 150km further south protects another set of pinnacles as the Cod Grounds Marine Park. These two tiny fishing exclusion zones were originally declared to protect threatened Grey Nurse Shark populations, and remain the only highly protected Australian Marine Park zones on the New South Wales continental shelf. Two much larger reefs (Elizabeth and Middleton) rise up from a submarine ridge in the Lord Howe Marine Park, 550km to the east. Even further to the east – much closer to New Caledonia and New Zealand than to Australia – sits the Norfolk Marine Park, which extends out from Norfolk Island's shores.

Solitary Islands Marine Park

Two adjoining and interlinked Solitary Islands Marine Parks exist off the northern NSW coast, one in New South Wales (NSW) waters managed by NSW Department of Primary Industries and the other by Parks Australia, an agency of the Commonwealth government. The inshore Solitary Islands Marine Park (in NSW waters) shelters flora and fauna within 3 nautical miles of the coast and shores of nearby islands. It is surrounded by an Australian Marine Park (in Commonwealth waters). The Yaegl people have had their native title rights recognised over parts of the NSW Solitary Islands Marine Park.

Recreational and restricted commercial fishing are allowed through most of the Solitary Islands Marine Park; however, the most unusual geographic feature – Pimpernel Rock – is highly protected within a National Park Zone, where fishing has been prohibited since 2001. Pimpernel Rock is a popular dive location due to its submerged set of three pinnacles that coalesce at the base and rise from 40m water depth to within 7m of the surface. It is an aggregation site for critically endangered Grey Nurse Sharks (*Carcharias taurus*) and protected as a National Park Zone to provide habitat for this migratory species during movement up and down the NSW and Queensland coasts. A huge cave undercuts the main pinnacle at 30 metres depth, also attracting the protected black cod (*Epinephelus daemelii*) and large schools of fish that shelter inside. Pelagic fishes, marine mammals, and turtles also visit.

The reef around Pimpernel Rock shallower than 18m depth is predominantly bare rock studded with barnacles and sea urchins, with occasional patches of Cunjevoi (*Pyura stolonifera*), encrusting sponges and filamentous algae. Cunjevoi are more abundant below 18m depth, as are hydroids and some gorgonians. Below 30m, the seabed is dominated by the stalked ascidian *Pyura spinifera*, sponges and sea whips. Sponge gardens extend out from the base of the three pinnacles.

Elsewhere in the Solitary Islands Marine Park, the ocean floor topography includes sand and flat bedrock interspersed with steep slopes and deeply scoured gutters but is generally too deep for recreational diving.

Zone Name	Zone Area (km²)
National Park Zone (no-fishing)	2
Multiple Use Zone (recreational and commercial fishing)	37
Special Purpose Zone (recreational and commercial fishing)	114

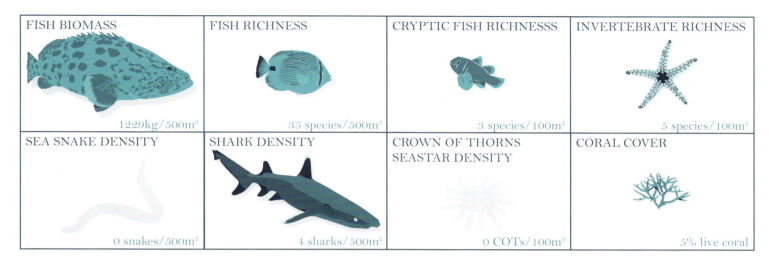

FISH BIOMASS	FISH RICHNESS	CRYPTIC FISH RICHNESSS	INVERTEBRATE RICHNESS
1229kg/500m²	35 species/500m²	3 species/100m²	5 species/100m²
SEA SNAKE DENSITY	SHARK DENSITY	CROWN OF THORNS SEASTAR DENSITY	CORAL COVER
0 snakes/500m²	4 sharks/500m²	0 COTs/100m²	5% live coral

RICK STUART-SMITH

RICK STUART-SMITH

Above: Below 30m depth, the near vertical walls of Pimpernel Rock carry a sparse cloak of seawhips, ascidians, sponges and encrusting algae. Anemones climb up the seawhips to reach stronger current flow. Schools of Mado (*Atypichthys strigatus*), Common Sweep (*Scorpis lineolata*) and other planktivorous fishes hover a few metres off the pinnacle wall.

Above: The entrances of deep caves at Pimpernel Rock are filled with schools of basslets. Redstripe Basslet (*Pseudanthias fasciatus*), Pimpernel Rock.

Below: Pennantfish (*Alectis ciliaris*), Pimpernel Rock, Solitary Islands Marine Park.

Above: Large Blue Groper (*Achoerodus viridis*) reside at Pimpernel Rock.

Below: A succession of predatory pelagic fish passes divers visiting Pimpernel Rock. Longnose Trevally (*Carangoides chrysophrys*).

Cod Grounds Marine Park

The Cod Grounds Marine Park is located in Commonwealth waters approximately 8km off the small town of Laurieton, New South Wales. The park encompasses three underwater pinnacles rising to about 18m depth from a seabed 40m deep. The complex seabed structure within the Marine Park forms a variety of distinctive habitats, including steep outcrops, shallow gutters, boulder/cobble slopes, and sand expanse[40].

The highly complex habitats of the Cod Grounds interact with the East Australian Current (EAC), the largest ocean current along the east coast of Australia. The EAC flows southwards in a series of eddies and areas of upwelling, drawing nutrient-rich water from a depth of at least 200m. EAC transportation of warmer waters and tropical larvae, and interactions with the complex seabed topography, promote strong currents and a rich flora and fauna within the Marine Park[41,42].

Fishing was banned in the Cod Grounds primarily to safeguard a critical aggregation site for threatened Grey Nurse Sharks (*Carcharias taurus*), which are common in this marine park.

This Marine Park is highly protected as a National Park Zone, with no fishing allowed and where all fishing gear must be stowed. It provides core habitat for Grey Nurse Sharks, which are protected as a threatened species, and which aggregate in large numbers in deep sand gutters between the reef pinnacles.

RLS surveys of the Cod Grounds and Solitary Islands Marine Parks revealed temperate species to be dominant in deeper areas, with a larger component of subtropical species in the warmer shallows. The shallowest benthic communities were typical of urchin barrens found elsewhere in NSW, with large numbers of the barrens-forming urchin *Centrostephanus rodgersii*[43]. Fish density is extremely high, due primarily to the high abundance of large fishes that are captured and reduced in numbers elsewhere. Grey Nurse Sharks are common.

Zone Name	Zone Area (km²)
National Park Zone (no-fishing)	4

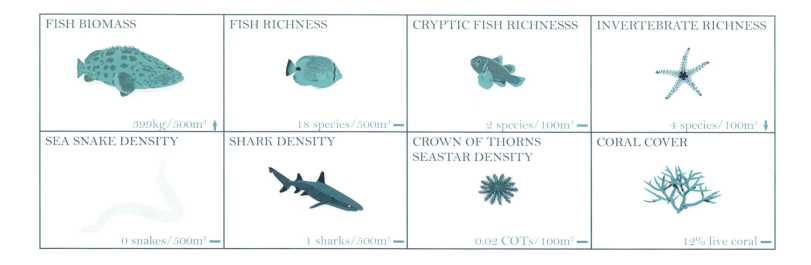

FISH BIOMASS	FISH RICHNESS	CRYPTIC FISH RICHNESSS	INVERTEBRATE RICHNESS
399kg/500m² ▲	18 species/500m² —	2 species/100m² —	4 species/100m² ▼
SEA SNAKE DENSITY	SHARK DENSITY	CROWN OF THORNS SEASTAR DENSITY	CORAL COVER
0 snakes/500m² —	1 sharks/500m² —	0.02 COTs/100m² —	12% live coral —

Below: Reefs below 30m in the Cod Grounds Marine Park echo an artist's multi-coloured palette. A rainbow of sponges, bryozoans, sea tulips and coralline algae spatter the deep reef.

RICK STUART-SMITH

Left: Predatory fishes are attracted to submarine pinnacles in part because of dense localised schools of smaller plankton-feeding fishes. Amberjack (*Seriola dumerili*), Cod Grounds Marine Park.

Right: A school of Nannygai (*Centroberyx affinis*) hovering by a Cod Grounds pinnacle.

Left: The Maori Rockcod (*Epinephelus undulatostriatus*) population is confined to the east coast of Australia. Cod Grounds Marine Park.

Right: Speckled anemones climb high above the seabed to capture plankton prey from a seawhip perch. Cod Grounds Marine Park.

Left: Due to high population numbers, not much searching is needed to stumble on Spotted Wobbegong (*Orectolobus maculatus*) camouflaged on the seabed. Cod Grounds Marine Park.

Right: Duncker's Pipehorse (*Solegnathus dunckeri*) generally lives in deep water (to 140m) and is very rarely seen by divers, but it is occasionally observed in the Cod Grounds Marine Park.

Above: Undercuts and caves within the Cod Grounds Marine Park are often occupied by fishes. Green Moray (*Gymnothorax prasinus*; left). Beardie (*Lotella rhacinus*; right).

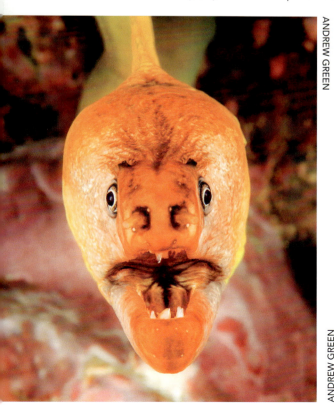

Lord Howe Marine Park

The Lord Howe Marine Park lies 500km east of the central New South Wales coast. This marine park is very large, extending nearly 700km north to south and 160km east to west. It drops from sea level to 6,000m depth and covers a total area of 110,126km². Lord Howe Island and Balls Pyramid sit in the centre of the park, but are excluded because they are in NSW State waters, and Elizabeth and Middleton Reefs are included in the north[44].

Elizabeth and Middleton Reefs are atoll-like platform reefs associated with a seamount chain that extends northward from Lord Howe Island[45]. They are the southernmost platform reefs in the world. Both reefs consist of an extensive lagoon

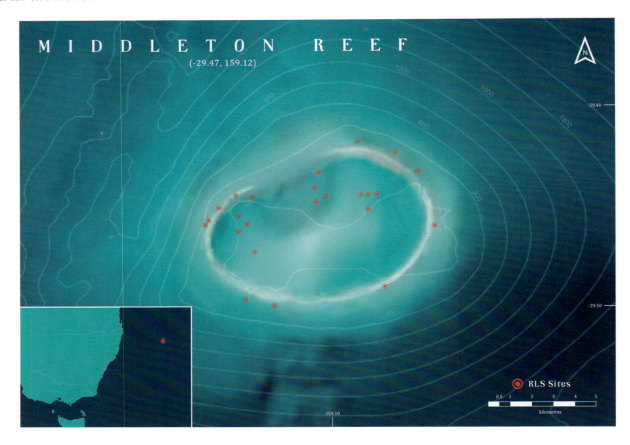

surrounded by a well-defined reef crest with characteristic spur-and-groove formations, and large passages in the north-west channelling water from the lagoon to the sea[46]. Prevailing winds from east to south-west result in very exposed reef front habitats on the southern face of the reefs[47].

Oceanographically, Elizabeth and Middleton Reefs generally receive warm water from eastward-flowing eddies of the East Australian Current[42] – a contrast to Lord Howe Island, which is primarily influenced by colder Tasman Sea water. East Australian Current eddies also intermittently transport larvae from the

212

Great Barrier Reef and Coral Sea. NSW coastal ecosystems to the west are a potential additional larval source for both temperate and subtropical species.

Elizabeth, Middleton and Lord Howe Island reefs host a unique and highly diverse assemblage of tropical and temperate organisms. Middleton Reef is highly protected within a Marine National Park Zone, while limited fishing is allowed in Elizabeth Reef's Recreational Use Zone. Marine park zoning around Lord Howe Island is more complex; outside of the NSW-managed Lord Howe Island Marine Park, the island is surrounded by a Habitat Protection Zone, with a small no-fishing zone to the east of the island, and another encompassing waters south of Ball's Pyramid.

Elizabeth and Middleton Reefs are together listed as Wetlands of International Importance under the Ramsar Convention. The outstanding environmental values recognised for the Convention include[48]:

1. They are representative of a unique ecosystem in the bioregion – the southernmost open-ocean coral reef platform in the world;
2. They support threatened species (Green Turtle, *Chelonia mydas*);
3. They support regionally high species diversity of (i) fishes, (ii) corals, (iii) molluscs, and (iv) sea cucumbers;
4. They support animal taxa at a vulnerable stage of their lifecycle (Galapagos Shark, *Carcharinus galapagensis*); and
5. They support the last known large population of Black Cod (*Epinephelus daemelii*).

Zone Name	Zone Area (km²)
National Park Zone (no-fishing)	9,273
Recreational Use Zone (limited access for recreational use)	1,170
Habitat Protection Zone (recreational and limited commercial fishing)	60,021
Multiple Use Zone (recreational and commercial fishing)	39,661

FISH BIOMASS	FISH RICHNESS	CRYPTIC FISH RICHNESSS	INVERTEBRATE RICHNESS
120kg/500m² —	34 species/500m² —	5 species/100m² ↑	6 species/100m² —
SEA SNAKE DENSITY	**SHARK DENSITY**	**CROWN OF THORNS SEASTAR DENSITY**	**CORAL COVER**
0 snakes/500m² —	2 sharks/500m² —	0.02 COTs/100m² —	24% live coral —

Above: Thin channels draining the reef in a spur-and-groove arrangement are evident from the air, extending from breakers to the sand fringe off south-western Middleton Reef.

Below: Elizabeth and Middleton Reefs are particularly hazardous to shipping because they lie beside key routes from eastern Australia to Asia, to the American west coast, and to Pacific Island ports. The first known shipwreck in the region was the 300-ton whaler *Britannia*, which ran onto Elizabeth Reef in 1806 while sailing from California to Sydney. At least 30 vessels are known to have been wrecked on the reefs between 1806 and 1972, 17 on Middleton Reef and 13 on Elizabeth Reef, but the true number is much higher. The most conspicuous remaining wreck is the *Runic*, which ran aground in a cyclone in 1961 on western Middleton Reef, and has now broken apart. This wreck was used for military target practice, resulting in unexploded bombs that remain a potential threat when diving nearby.

TONI COOPER

Above: Despite its remote isolation, corals at Middleton Reef are affected by region-wide heatwaves and associated bleaching events.

Right: Galapagos Sharks (*Carcharhinus galapagensis*) are observed on most dives at Middleton Reef, a refuge for the species.

Below left: Pelagic fish predators frequently glide past during dives along the reef edge. Thicklip Trevally (*Carangoides orthogrammus*), Middleton Reef.

Below right: Schools of Silver Trevally (*Pseudocaranx dentex*) forage amongst sand and rubble in Middleton Reef lagoon.

Above: Southern Elizabeth Reef.

Above left: Juvenile Cleaner Wrasse (*Labroides dimidiatus*) tending a Blacktip Rockcod (*Epinephelus fasciatus*). Middleton Reef.

Above right: Common Lionfish (*Pterois volitans*). Middleton Reef.

Above left: Sea hares (*Aplysia dactylomela*) are amongst the largest invertebrate algal grazers. Middleton Reef lagoon.

Above right: Imperial Hermit Crab (*Calcinus imperialis*) amongst coral. Middleton Reef.

Below: Middleton Reef lagoon, with dendritic channels leading to north-western entrance lagoon.

ELIZABETH REEF
(-29.94, 159.06)

RLS sites

Undulate Moray (*Gymnothorax undulatus*) moves out from cover at night to search for prey in Elizabeth Reef lagoon.

Little Hooded Triplefin (*Helcogramma chica*) blends amongst algal turf, Elizabeth Reef.

Above: Huge fish schools add a degree of difficulty when completing fish surveys. Elizabeth Reef.

Below: Reef edge soft and hard corals, with passing Threeband Butterflyfish (*Chaetodon tricinctus*), a species largely confined to the Lord Howe region. Elizabeth Reef.

The icon of Elizabeth and Middleton Reefs is the Black Cod (*Epinephelus daemelii*). Populations of this large grouper had catastrophically declined along the Australian mainland by 1980. Consequently fishing was prohibited on the two reefs in 1987 to sustain the species in its last stronghold. Regulations were relaxed on Elizabeth Reef in 2004 to allow recreational fishing for other species. Elizabeth Reef.

Opposite: Barnacles attached to floating log. *Lepas anserifera*, Elizabeth Reef.

Left: Nudibranchs use bright coloration to highlight the presence of toxic defence chemicals in their bodies. These chemicals are obtained from sponge and soft coral food. Jackson's Nudibranch (*Hypselodoris jacksoni*), Elizabeth Reef.

Right: All nudibranchs have both male and female organs, frequently forming mating pairs or larger groups to exchange sperm. Tryon's Nudibranch (*Hypselodoris tryoni*), Elizabeth Reef.

TONI COOPER

Above: Grazing fishes often occur in large schools. This provides defence from predators, and also mob access to small patches of algae that are actively cultivated and defended by solitary gardener fishes. Spotted Sawtail (*Prionurus maculatus*), Elizabeth Reef lagoon.

Above: Floating vegetation provides a dispersal pathway across deep ocean for some fishes and invertebrates. In addition to hosting juveniles of offshore species such as Speckled Driftfish (*Psenes cyanophrys*; bottom left), drifting logs transport reef species including juvenile Brassy Drummer (*Kyphosus vaigiensis*; bottom right) between islands. Elizabeth Reef.

Above: Blunt Slipper Lobster (*Scyllarides squammosus*), a night roving predator. Elizabeth Reef lagoon.

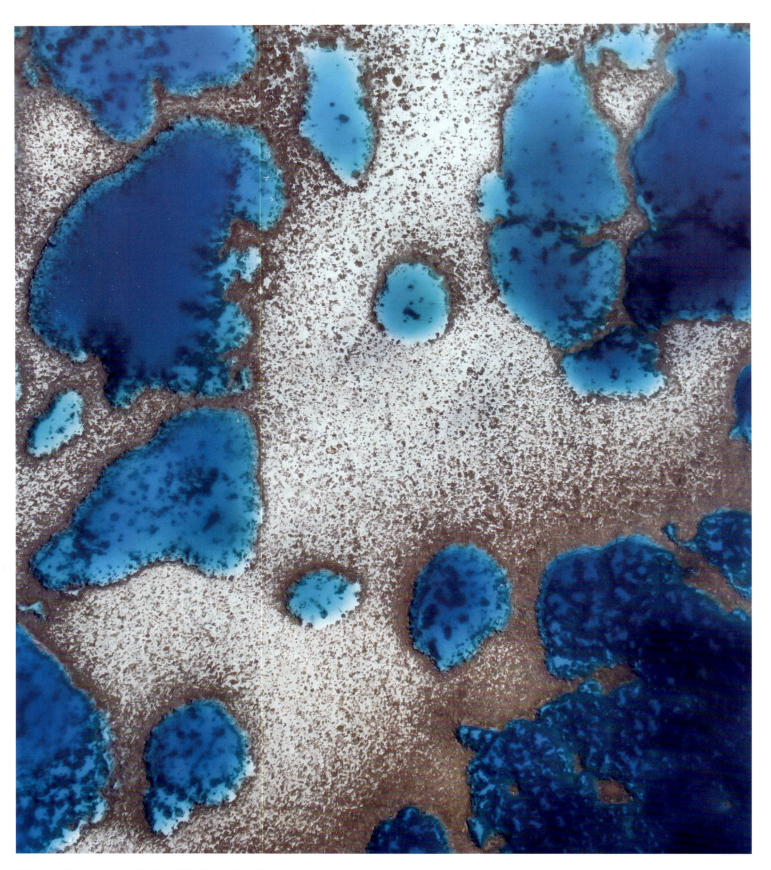

Above: Blue lagoon holes, Elizabeth Reef.

Norfolk Marine Park

NORFOLK
ISLAND
(-29.04, 167.95)

⊙ RLS Sites

The Norfolk Marine Park is located approximately 1,600km north-east of Sydney (New South Wales). With a span of about 700km, it covers an area of approximately 188,000km². The Norfolk Ridge – a long narrow undersea mountain range with pinnacles and seamounts – forms the spine of the park. This ridge provides underwater stepping stones, connecting the land masses (and marine life) between New Zealand and New Caledonia. The park encompasses depths ranging from 5,000m to the high-water mark, making it the only Australian Marine Park directly adjacent to a local human community and accessible from shore. A Special Purpose Zone surrounds Norfolk Island, allowing activities such as diving and fishing as long as they don't involve destructive techniques.

Archaeological evidence indicates that Norfolk Island was occupied by Polynesian seafarers between the 13th and 15th centuries. Captain James Cook was the first European visitor, sighting the island in 1774. The island was settled by the British in 1788, just five weeks after the First Fleet arrived in Sydney. It was chosen because the tall Norfolk Island pines, and the flax which grew there, were mistakenly

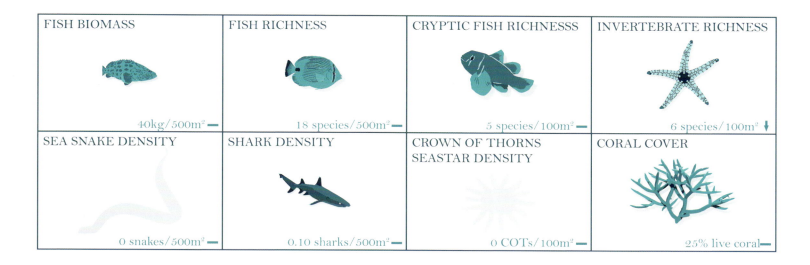

FISH BIOMASS	FISH RICHNESS	CRYPTIC FISH RICHNESS	INVERTEBRATE RICHNESS
40kg/500m² —	18 species/500m² —	5 species/100m² —	6 species/100m² ↓
SEA SNAKE DENSITY	SHARK DENSITY	CROWN OF THORNS SEASTAR DENSITY	CORAL COVER
0 snakes/500m² —	0.10 sharks/500m² —	0 COTs/100m² —	25% live coral —

identified as useful for ship's masts and sails. Nevertheless, the island's rich soil and mild climate created ideal conditions for farming, providing food for the growing colony in Sydney.

Convicts and free settlers made Norfolk Island their home until 1814, when the island was abandoned due to its isolation and lack of safe landing sites. A second convict settlement began in 1825, with Norfolk Island becoming infamous for the harsh treatment prisoners received. The settlement was wound down in the 1850s, after convict transportation to New South Wales ceased.

Settlement recommenced from 1856, when the island was given to Pitcairn Island descendants of the infamous *Bounty* mutineers and their Tahitian wives. The 196 new arrivals received 50-acre land grants. Clearing of much of the native rainforest over the next century for forestry, fruit farms and cattle pasture converted the landscape into the pastoral countryside of rolling green hills that we see today.

The oceanography and ecology of the region are strongly influenced by the South Equatorial Current, which carries tropical Pacific waters towards the Coral and Tasman Seas. The shallow reefs of Norfolk Marine Park have developed on the southern boundary of coral reef formation, supporting a mix of tropical, temperate and endemic flora and fauna. Reef communities are further structured by gradients in wave exposure around Norfolk Island. A combination of prevailing swells, winds from multiple directions, and few enclosed bays, results in moderately to strongly wave-exposed reef habitats, with only a small shallow lagoon at Emily Bay in the south supporting a sheltered coral reef habitat. The isolation of Norfolk Marine Park reefs from other shallow reefs in the subtropical zone has contributed to the presence of a number of regional endemic species, and high abundances of some species that are rare or unusual elsewhere[49].

Zone Name	Zone Area (km²)
National Park Zone (no-fishing)	41,661
Habitat Protection Zone (recreational and limited commercial fishing)	138,796
Special Purpose Use Zone (recreational and most commercial fishing)	7,986

North-eastern coast, Norfolk Island.

'Located in the Pacific Ocean between New Zealand and New Caledonia, a 2.5-hour flight from Australia, and formed from several volcanic eruptions, the ancient cliffs of Norfolk Island have 32km of very steep and rugged coastline that were formed between 3.1 million and 2.3 million years ago. Like stepping back in time (or on 'island time') Norfolk Island is a culturally-rich, friendly and really beautiful island, where wandering cows have right of way, Norfolk Pines are in every view, and everyone has time to stop and chat. The vistas from the sheer cliffs and lookouts provide an endless expanse of inky blue ocean, with foaming white-water swirling against the rugged Jurassic cliffs below. If above water is gorgeous, then below water is even more breathtaking – the crystal-clear waters providing an array of diverse and colourful coral and algal dominated reefs with a diversity of fish and marine life.'

SALLYANN GUDGE, RLS BLOG.

Western coast, Norfolk Island.

'The offshore reefs welcomed us with clear warm (23–24°C) water, where the myriad of hard and soft corals, mixed with colourful macro-algae diversity were a feast for us all. The fish life was abundant and diverse, with many similar species to those found within the Lord Howe Island Marine Park some 900km to the south-west.'

SALLYANN GUDGE, RLS BLOG.

Wideband Anemonefish (*Amphiprion latezonatus*) is a characteristic species of the Lord Howe/Norfolk region, only occurring elsewhere as vagrant waifs. Norfolk Marine Park.

Below left: The Notch-head Marblefish (*Aplodactylus etheridgii*) is a large herbivorous fish restricted to the region, with closest relatives on temperate Australian and New Zealand kelp-covered reefs. Norfolk Marine Park.

Below right: Smoky Puller (*Chromis fumea*), Norfolk Marine Park.

Above right: Reefs around the Norfolk Island coast often drop quickly onto sand. Norfolk Marine Park.

'*Launching boats on Norfolk Island involves a derrick or hydraulic crane rather than boat ramp, and it surely is an operation to behold! Luckily, the lagoon sites (Emily Bay and Slaughterhouse Bay) provide protected options around the low tide, although many days the swell was so large that it was breaking over the fringing reef. Outside the small lagoon there are two larger uninhabited offshore islands – Phillip and Nepean – and several smaller islets, where beautiful diverse coral reef and algal communities exist beneath the surging waves.*'

SALLYANN GUDGE, RLS BLOG.

Opposite above: Dusky Shark (*Carcharhinus obscurus*), Norfolk Marine Park.

Opposite middle: Banded Scalyfin (*Parma polylepis*) cultivate gardens of seaweed that they vigorously defend. Without hesitation, they will confront divers who cross into their territories.

Opposite below: Shallow reef habitats throughout Norfolk Marine Park comprise a mosaic of coral, seaweed and sponge micro-habitats.

Small patch reefs (*Turbinaria mesenterina*; above) and coral heads (*Euphyllia ancora*; below) in Emily Bay lagoon.

Above: Coastal reef, Emily Bay, Norfolk Marine Park.   Below: Emily Bay, Norfolk Island south coast.

The sandy seabed in Emily Bay supports a localised community of soft-sediment dwellers, including the Grinning Tun (*Malea ringens*), a predator of sea cucumbers.

Wideband Anemonefish (*Amphiprion latezonatus*), Norfolk Marine Park.

Above: The gaping mouth of the Grey Moray (*Gymnothorax nubilis*) – a regional endemic species – may appear threatening, but is for breathing in water. Norfolk Island.

Above: Reef Life Survey diver, Norfolk Island.

Opposite: Green Moon Wrasse (*Thalassoma lutescens*) scoot continuously across shallow coral reefs, often oblivious to divers. Norfolk Marine Park.

Left: Yellowtail Kingfish (*Seriola lalandi*) school on outer Norfolk Island reef.

Right: Silver Trevally (*Pseudocaranx dentex*), Norfolk Island.

Above left: Norfolk Island reefscape, with Mado (*Atypichthys latus*) and Norfolk Cardinalfish (*Apogon norfolcensis*).

Above right: Table coral (*Acropora* species), Norfolk Island.

Below left: Orange Long-armed Sea Star (*Ophidiaster confertus*), Norfolk Island.

Below right: Schooling Striped Catfish (*Plotosus lineatus*) move in a tight ball while taking turns to feed on the seabed. Norfolk Island.

Above: Diver surveying the underside of a giant Norfolk Island underwater arch.

Below: Table corals shelter fishes from predators and, in shallow water, also provide shade from damaging effects of ultraviolet rays. Norfolk Cardinalfish (*Ostorhinchus norfolcensis*), Norfolk Island.

South-east
Marine Parks
Network

A total of 14 Australian Marine Parks distributed around the south-eastern continent from Kangaroo Island (South Australia) to the Victorian/New South Wales border, and off Tasmania and subantarctic Macquarie island, form the South-east Network. Regulations associated with this 388,464km² network were first implemented in 2007. The network covers waters ranging from 50m to over 6,000m in depth. It includes seamounts up to 4,000m in height that arise from the continental slope and abyssal plain. Seamounts act as productive oases of life by intensifying deep current flow and redirecting nutrients to the sea surface. The network additionally includes submarine canyons that cut through the continental shelf, underwater mountain chains and plateaus, and part of the ancient lake depression at the centre of Bass Strait.

Aboriginal people of south-eastern Australia have been using and managing their Sea Country for tens of thousands of years, including before rising sea levels created these marine environments. Bass Strait and elsewhere on the inner continental shelf were dry land 18,000 years ago, a time when the sea level was 120m lower than now. Rising sea levels and coastal flooding greatly affected local Aboriginal communities as conditions warmed with the melting of glaciers. This period of migration landward to higher ground is recorded in oral histories.

Marine life is diverse and also characterised by a very high proportion of species found nowhere else on Earth. This temperate flora and fauna has evolved in isolation through 55 million years since the Australian continent separated from Antarctica. Over 80 per cent of plants and animals in southern Australia are restricted to this continent[50].

South-east Australian Marine Parks are all located at considerable distance from coastal population centres. Some boat-based tourism is associated with the Bonney Upwelling, which extends along the

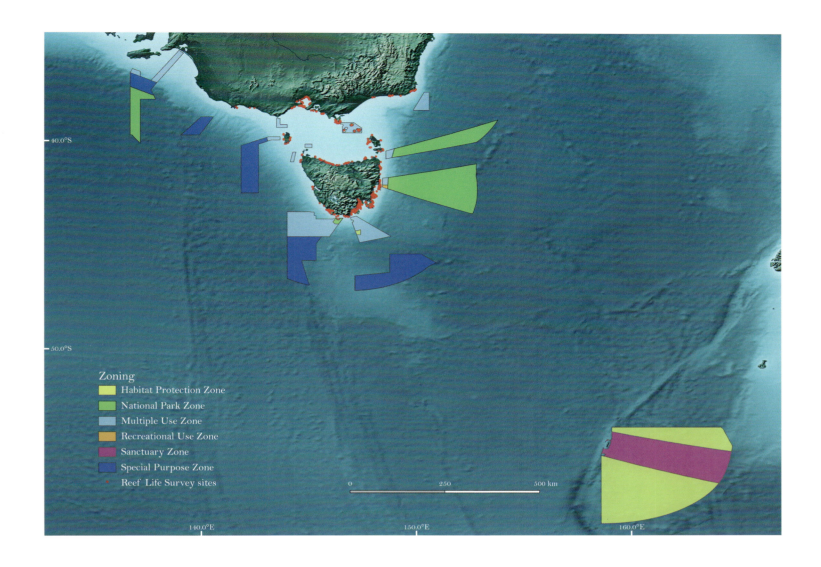

Zoning
- Habitat Protection Zone
- National Park Zone
- Multiple Use Zone
- Recreational Use Zone
- Sanctuary Zone
- Special Purpose Zone
- Reef Life Survey sites

narrow continental shelf from Robe (South Australia) to Portland (Victoria). It is the largest and most predictable upwelling of nutrient-rich water off Australia. Nutrients channelled through slender shelf canyons support phytoplankton blooms, which fuel large populations of krill (*Nyctiphanes australis*), sardines (*Sardinops sagax*) and anchovies (*Engraulis australis*), which in turn attract large numbers of penguins, flying seabirds, tuna, fur seals, and whales. Around 100 blue whales (*Balaenoptera musculus*) visit the Bonney Upwelling area every year.

Few dive opportunities exist in marine parks within the South-east Network due to extreme water depths. Relatively shallow seabed features are located within the Freycinet Marine Park, off Tasmania's east coast. This is the only Marine Park within the South-East network where recreational diving has occured that we are aware of, but required the use of specialised technical diving equipment.

Freycinet Marine Park

The Freycinet Marine Park extends off the Tasmanian coast as a pie slice expanding from Bicheno out to the edge of Australia's Exclusive Economic Zone. Fishing is excluded from offshore waters in the marine park deeper than about 1,000m; recreational fishing and many forms of commercial fishing are permitted on the continental shelf and slope. The Freycinet Marine Park was declared in part to protect deep seamounts rising from the abyssal plain that had not been fished, and to safeguard important foraging and migration areas for birds such as Wandering, Black-browed and Shy Albatrosses, Cape Petrel and Fairy Prion, and cetaceans such as Humpback, Southern Right and Sei Whales and Orca.

The most accessible dive site in the Freycinet Marine Park is known to local fishers as 'Joe's Reef', an unusual 200m-long granite reef surrounded by sandy substrate. This reef is located 11km off the coast between 59–83m depth. It has recently been explored and documented by deep-diving citizen scientists. They discovered a remarkable community – large numbers of tree-forming black corals amongst spectacularly diverse gardens of sponges, soft corals and bryozoans. Deeper areas around the reef margins included sand, boulders, cobbles and low relief reef.

Zone Name	Zone Area (km²)
Marine National Park Zone (no-fishing)	56,793
Recreational Use Zone (limited access for recreational use)	323
Multiple Use Zone (recreational and commercial fishing)	825

ALL PHOTOS JAMES PARKINSON

As described below by James Parkinson, diving the deep reefs of the Freycinet Marine Park poses many technical challenges to cope with conditions that are well beyond those experienced in shallow waters:

'Diving Joe's reef is considered a technical dive because of the specialised gear and training required to safely dive the site. A technical dive is a planned dive, with specialist equipment, beyond the limits of recreational diving where the diver generally can't ascend directly to the surface due to an overhead environment or a decompression ceiling, as is the case when diving Joe's.

The equipment typically used to dive Joe's consists of a closed-circuit rebreather unit, where the diver's exhaled breath is recycled in a closed loop, as opposed to release into the water as happens with standard SCUBA gear. Open-circuit bailout cylinders are used to get you back to the surface safely should the rebreather fail, which thankfully is very rare.

A dive on Joe's can range in depth from 60–80m with the dive requiring specialised gas mix called trimix which contains helium to offset the effects of nitrogen narcosis, also known as 'rapture of the deep', a debilitating condition if breathing normal air at this depth. The gas mix has a reduced oxygen content, typically 15 per cent, as increased oxygen approaching that of air can be toxic at those depths. Typical dive time on Joe's range from 20–30 minutes resulting in 60–90 minutes of decompression, basically three minutes of decompression for every minute on the bottom.

JAMES PARKINSON

Decompression is spent drifting in blue water on a decompression station, deployed tethered to the shot line then released on ascent. One of the main challenges to overcome when diving Joe's is the current. Dives are planned to coincide with good weather (light winds and low swell) but the current is unpredictable and a big unknown on any given day. We don't know what the current will be doing until we get out to the site and deploy a shot line, a line with an anchor and buoy that is placed on the dive location that divers descend and ascend. Sometimes the shot anchor won't hold and drifts off the site quickly, and when it does hold the buoy can be pulled underwater by the strong currents.'

A Draughtboard Shark (*Caphaloscyllium laticeps*) has concealed its egg case within black coral branches. The yellow embryo is developing inside.

Reef habitat in the Freycinet Marine Park is confined to deep waters with inadequate sunlight reaching the seabed for seaweeds to grow. Consequently, without algae taking up space, the reef hosts a diverse community of sponges, bryozoans and ascidians. This is particularly the case in current-swept locations where filter-feeding invertebrates sit in a passing stream of planktonic food particles. Plankton-feeding fishes, such as this flock of Butterfly Perch (*Caesioperca lepidoptera*), can be hyper-abundant in high current areas. Black coral colonies (*Antipathes* species) provide additional 2m-high structure to the seabed, creating a micro-oasis.

Older branches at the base of black corals provide a substrate for colonisation by other seabed invertebrates, including dense creeping clusters of pink jewel anemones (*Corynactis* species).

Detail of black coral branch, showing individual polyps. While black coral is named for the colour of the species' skeletons, they can have brightly coloured living tissue.

Black corals grow to over 4 m on Joe's Reef, an outstanding size that places them amongst the largest invertebrates in southern Australia.

Macquarie Island Marine Park

Macquarie island Marine Park lies 1,500km south-east of Tasmania at 55°S, approximately halfway between Australia and Antarctica. It includes the subantarctic waters to the east and south of Macquarie Island Marine Reserve, a Tasmanian marine park that extends 3 nautical miles from the Macquarie Island shoreline. The marine park was originally proclaimed in 1999 as the Macquarie Island Marine Park, with a name change in 2007 to the Macquarie Island Commonwealth Marine Reserve, then back again to Macquarie Island Marine Park in 2017.

The Macquarie Island region was inscribed on the World Heritage List in 1997 because of its outstanding natural values. It is the only place on Earth where rocks from the Earth's mantle (6km below the ocean floor) have been pushed by tectonic forces above sea level. The island was formed as the exposed crest of the Macquarie Ridge, with undersea canyons, escarpments, abyssal hills, ridges, slopes and trenches nearby in the Macquarie Island Marine Park.

Macquarie Island supports a remarkable concentration of wildlife. The island provides an important breeding area for Southern Elephant Seals; Antarctic, Subantarctic and New Zealand Fur Seals; Royal, King, Southern Rockhopper and Gentoo Penguins; Light-mantled Sooty, Grey-headed, Black-browed and Wandering Albatrosses; and many burrow-nesting seabirds including petrels. Some breeding colonies are enormous, numbering from tens to hundreds of thousands of individuals. Macquarie Island Marine Park supports these breeding concentrations by safeguarding the offshore foraging areas for parents stocking up on food to feed their young, and for juveniles once they migrate offshore.

Rocky reef in the Macquarie Island Marine Park is too deep for recreational diving, with the seabed ranging from 86m to 6,341m depth. Regardless, interactions between mammals, birds, fishes and krill in surface waters never cease to enthral the few visitors.

Zone Name	Zone Area (km²)
Sanctuary Zone (no-fishing)	58,000
Habitat Protection Zone (recreational and limited commercial fishing)	104,000

More than 150,000 King Penguins (*Aptenodytes patagonicus*) breed in Lusitania Bay, Macquarie Island. Their huge demand for food is partly supplied by foraging in Macquarie Island Marine Park waters, located 3 nautical miles directly offshore.

Most marine species found off Macquarie Island have distributions that extend around the Southern Ocean.

Right: Mawson's Sea Star (*Anasterias mawsoni*) and periwinkles (*Cantharidus capillaceus*) traverse green felt algae (*Codium subantarcticum*) in inshore Macquarie Island waters.

Below: The Magellanic Icefish (*Paranotothenia magellanica*) is the most abundant fish species seen around Macquarie Island, and is a major food source for many marine mammal and seabird species.

Bottom: Southern Elephant Seals (*Mirounga leonina*) regularly search for food below 1,000m depth in the Macquarie Island Marine Park, each dive lasting up to two hours[51]. As the largest marine mammal apart from whales, they have massive dietary requirements. Macquarie Island.

Above: The Light-mantled Sooty Albatross (*Phoebetria palpebrate*) spends its life in offshore waters other than when nesting. Macquarie Island.

Right: Three fur seal species forage in Macquarie Island Marine Park waters: New Zealand (*Arctocephalus forsteri*), Subantarctic (*Arctocephalus tropicalis*) and Antarctic (*Arctocephalus gazella*). The range of the New Zealand Fur Seal (shown at Macquarie Island) extends north to other marine parks in the South-east Network.

Below: The Macquarie shag (*Leucocarbo purpurascens*), a blue-eyed seabird endemic to Macquarie Island.

Acknowledgements

This Our Marine Parks Grants project received grant funding from the Australian Government.

The views and opinions expressed in this publication are those of the authors.

Field surveys and analyses were made possible by generous support provided by The Ian Potter Foundation, Minderoo Foundation, and Parks Australia.

Reef Life Survey would also like to thank Whitsunday Escape and the many people who assisted field trips on board sailing vessel *Eviota*.

RLS divers: Andrew Green, Graham Edgar, Ian Shaw, Sue Baker, Stuart Kininmonth, Antonia Cooper, Tim Crawford, Bill Barker, Bob Edgar, Anna Edgar, German Soler, Don Love, Garrick Smith, Jamie Hicks, Jennifer Hine, John Turnbull, Kim Sebo, Marlene Davey, Sallyann Gudge, Sam Griffiths, Tyson Jones, Sue Newson, Kevin Smith, Alicia Sutton, Ben Jones, Margo Smith, Ashley Smith, Anna Creswell, Rebecca Watson, Cheryl Petty, Marjon Phur, Rick Stuart-Smith, Michael Brooker, Lotte Rivers, Mike Sugden, Joe Shields, Tanja Ponudic, Nestor Echedey Bosch Guerra, Russell Thomson, Beth Strain, Brian Busteed, Tom Davis, Nicola Davis, Meryl Larkin, Nick Mooney, Scott Ling, Yanir Seroussi, Caitlin Woods, Tim Alexander, Sophie Edgar, Carly Giosio, John Lemburg, Laura Smith, Kate Tinson, Charlie Bedford, Martin Filleul, German Soler and Simon Talbot.

Skippers: Derek Shields, Graham Edgar, Ian Donaldson, Sam Griffiths, Mark Shepherd, David Mason, Pieter van der Woude, Graeme Ewing, Mike Sugden, Mitch Graham and James Edwards.

Infographics: Ella Clausius and Freddie Heather.

Technical advice and support: Christine Fernance, Ella Clausius, Freddie Heather, Lizzie Oh, Neville Barrett, Jac Monk, Justin Hulls, Emre Turak, Phil Alderslade and Parks Australia staff.

Maps: Jo Hannah Aestre and Antonia Cooper.

Image editing: Tasha Waller.

Quoted text: James Parkinson, Ian Shaw, Bill Barker, Kirsty Whitman, Sallyann Gudge, Paul Day, Nestor Echedey Bosch Guerra, Caitlin Kuempel and Kevin Smith.

Photographers: Graham Edgar (all uncredited photographs), Rick Stuart-Smith, Antonia Cooper, Andrew Green, Ian Shaw, James Parkinson, Kevin Smith, German Soler and Bob Edgar.

Reef Life Survey thanks the Gumurr Marthakal Rangers, Dhimurru Rangers, Carpentaria Land Council Aboriginal Corporation, and fisheries officer Klaus Jeffery for facilitating and participating in visits to Sea Country.

References

1   Bennett, S. *et al.* The 'Great Southern Reef': Social, ecological and economic value of Australia's neglected kelp forests. *Marine and Freshwater Research* 67, 47–56, doi:10.1071/MF15232 (2016).

2   Stuart-Smith, J. *et al.* Conservation challenges for the most threatened family of marine bony fishes (handfishes: Brachionichthyidae). *Biological Conservation* 252 doi.org/10.1016/j.biocon.2020.108831 (2020).

3   Stuart-Smith, R.D. *et al.* Loss of native rocky reef biodiversity in Australian metropolitan embayments. *Marine Pollution Bulletin* 95, 324–332, doi:10.1016/j.marpolbul.2015.03.023 (2015).

4   Johnson, C.R. *et al.* Climate change cascades: shifts in oceanography, species' ranges and subtidal marine community dynamics in eastern Tasmania. *Journal of Experimental Marine Biology and Ecology* 400, 17–32 (2011).

5   Ling, S.D. *et al.* Pollution signature for temperate reef biodiversity is short and simple. *Marine Pollution Bulletin* 130, 159–169, doi:10.1016/j.marpolbul.2018.02.053 (2018).

6   Edgar, G.J., Samson, C.R. and Barrett, N.S. Species extinction in the marine environment: Tasmania as a regional example of overlooked losses in biodiversity. *Conservation Biology* 19, 1294–1300 (2005).

7   Silberstein, K., Chiffings, A.W. and McComb, A.J. The loss of seagrass in Cockburn Sound, Western Australia. III. The effect of epiphytes on the productivity of Posidonia australis Hook. f. *Aquatic Botany* 24, 355–371 (1986).

8   Jennings, S. and Kaiser, M.J. in *Advances in Marine Biology* 201–352 (1998).

9   Carlton, J.T. Man's role in changing the face of the ocean: biological invasions and implications for conservation of nearshore environment. *Conservation Biology* 3, 265–273 (1989).

10 Carlton, J.T. and Geller, J.B. Ecological roulette: the global transport of non-indigenous marine organisms. *Science* 261, 78–82 (1993).

11 Laist, D.W. and Overview of the biological effects of lost and discarded plastic debris in the marine environment. *Marine Pollution Bulletin* 18, 319–326 (1987).

12 Williams, R. *et al.* Impacts of anthropogenic noise on marine life: Publication patterns, new discoveries, and future directions in research and management. *Ocean and Coastal Management* 115, 17–24 (2015).

13 Dayton, P.K., Tegner, M.J., Edwards, P.B. and Riser, K.L. Sliding baselines, ghosts, and reduced expectations in kelp forest communities. *Ecological Applications* 8, 309–322 (1998).

14 Reef Life Survey Foundation. *Standardised survey procedures for monitoring rocky & coral reef ecological communities.* (Reef Life Survey Foundation, 2019).

15 Devillers, R. *et al.* Reinventing residual reserves in the sea: Are we favouring ease of establishment over need for protection? *Aquatic Conservation: Marine and Freshwater Ecosystems* 25, 480–504, doi:10.1002/aqc.2445 (2015).

16 Devillers, R. *et al.* Residual marine protected areas five years on: Are we still favouring ease of establishment over need for protection? *Aquatic Conservation: Marine and Freshwater Ecosystems* 30, 1758-1764, doi:10.1002/aqc.3374 (2020).

17 Plaisance, L., Caley, M.J., Brainard, R.E. and Knowlton, N. The diversity of coral reefs: What are we missing? *PLoS ONE* 6, e25026. doi:25010.21371/journal.pone.0025026 (2011).

18 Ceccarelli, D. M. *et al.* in *Advances in Marine Biology* Vol. 66 (ed M. Lesser) 213–290 (Academic Press, 2013).

19 Gilmour, J. *et al.* Data compilation and analysis for Rowley Shoals: Mermaid, Imperieuse and Clerke reefs. (Report to the Department of the Environment, Water, Heritage and the Arts by the Australian Institute of Marine Science, Perth, 2007).

20 Commonwealth of Australia. Recovery plan for the grey nurse shark (*Carcharias taurus*) in Australia., (Environment Australia, Canberra, 2001).

21 Edgar, G. *et al. Reef Life Survey Assessment of Coral Reef Biodiversity in the North-west Marine Parks Network.* (Reef Life Survey Foundation, 2020).

22 Baker, C., Potter, A., Tran, M. and Heap, A.D. *Geomorphology and sedimentology of the Northwest Marine Region of Australia. Geoscience Australia, Record 2008/07.*, (Geoscience Australia, 2008).

23 Skewes, T.D. *et al.* Survery and stock size estimates of the shallow reef (0–15m deep) and shoal area (15–50m deep) marine resources and habitat mapping within the Timor Sea MOU74 Box. Volume 2: Habitat mapping and coral dieback. (Report for the FRRF and Environment Australia by CSIRO Division of Marine Research, Canberra, 1999).

24 Rees, M., Colquhoun, J., Smith, L. and Heyward, A. Surveys of Trochus, Holothuria, giant clams and the coral

communities at Ashmore Reef, Cartier Reef and Mermaid Reef, Northwestern Australia: 2003. (AIMS Report Produced for the Department of Environment and Heritage, Townsville, 2003).

25 Ceccarelli, D.M., Richards, Z.T., Pratchett, M.S. and Cvitanovic, C. Rapid increase in coral cover on an isolated coral reef, the Ashmore Reef National Nature Reserve, northwestern Australia. *Marine and Freshwater Research* **62**, 1214–1220 (2011).

26 Veron, J.E.N. *A biogeographic database of hermatypic corals.* (Australian Institute of Marine Sciences, 1993).

27 Milton, D.A. Birds of Ashmore Reef National Nature Reserve: an assessment of its importance for seabirds and waders. *The Beagle, Records of the Museums and Art Gallery of the Northern Territory* **Suppl. 1**, 133–141 (2005).

28 Adhuri, D. in *Macassan History and Heritage: Journeys, Encounters and Influences* 183–203 (ANU Press, 2013).

29 Somaweera, R. *et al.* Pinpointing Drivers of Extirpation in Sea Snakes: A Synthesis of Evidence From Ashmore Reef. *Frontiers in Marine Science* **8**, doi:10.3389/fmars.2021.658756 (2021).

30 Brandl Simon, J., Goatley Christopher, H.R., Bellwood David, R. and Tornabene, L. The hidden half: ecology and evolution of cryptobenthic fishes on coral reefs. *Biological Reviews* **0**, doi:10.1111/brv.12423 (2018).

31 Duke, N.C. *et al.* Large-scale dieback of mangroves in Australia's Gulf of Carpentaria: a severe ecosystem response, coincidental with an unusually extreme weather event. *Marine and Freshwater Research* **68**, 1816–1829 (2017).

32 Sayre, P. *Australia's Coral Sea Islands & Marine Park.* ISBN: 978-0-6485821-0-6 (2019).

33 Edgar, G.J., Ceccarelli, D.M. and Stuart-Smith, R.D. *Assessment of coral reef biodiversity in the Coral Sea. Unpublished report to Parks Australia. Available at reeflifesurvey.com/publications/ assessment-of-coral-reef-biodiversity-in-the-coral-sea/* (Reef Life Survey Foundation, 2015).

34 Ceccarelli, D.M. *et al.* The Coral Sea. Physical Environment, Ecosystem Status and Biodiversity Assets. *Advances in Marine Biology* **66**, 213–290 (2013).

35 Kelly, L.W. *et al.* Black reefs: iron-induced phase shifts on coral reefs. *The ISME Journal* **6**, 638-649, doi:10.1038/ismej.2011.114 (2012).

36 Stuart-Smith, R. *et al. Reef Life Survey Assessment of Marine Biodiversity in Geographe Bay. Report to Parks Australia.* (Reef Life Survey Foundation Incorporated., 2020).

37 Edgar, G.J. Predator-prey interactions in seagrass beds. III. Impacts of the western rock lobster *Panulirus cygnus* George on epifaunal gastropod populations. *Journal of Experimental Marine Biology and Ecology* **139**, 33–42 (1990).

38 Keene, J., Baker, C., Tran, M. and Potter, A. Sedimentology and geomorphology of the East Marine region of Australia – A spatial analysis. (Record 2008/10. Geoscience Australia, Canberra, 2008).

39 Stuart-Smith, R.D., Ceccarelli, D., Edgar, G. J. and Cooper, A.T. *2016 Biodiversity surveys of the Cod Grounds and Pimpernel Rocks Commonwealth Marine Reserves. Report for Parks Australia, Department of the Environment.* (Reef Life Survey Foundation Incorporated, 2017).

40 Davies, P. *et al.* Cod Grounds Commonwealth Marine Reserve: Swath survey and habitat classification. (Final report to the Department of the Environment, Water, Heritage and the Arts, Canberra, 2008).

41 Oke, P.R. and Middleton, J.H. Nutrient enrichment off Port Stephens: the role of the East Australian Current. *Continental Shelf Research* **21**, 587–606 (2001).

42 Brewer, D.T. *et al.* Ecosystems of the East Marine Planning Region. (Report to Department of Environment and Water Resources by CSIRO, Cleveland, 2007).

43 Rule, M.J. and Smith, S.D.A. Depth-associated patterns in the development of benthic assemblages on artificial substrata deployed on shallow, subtropical reefs. *Journal of Experimental Marine Biology and Ecology* **345**, 38–51 (2007).

44 Edgar, G.J., Ceccarelli, D., Stuart-Smith, R.D. and Cooper, A.T. *Biodiversity surveys of the Elizabeth and Middleton Reefs Marine National Park Reserve, 2013 and 2018.* (Reef Life Survey Foundation Incorporated, 2018).

45 Woodroffe, C.D., Kennedy, D.M., Jones, B.G. and Phipps, C.V.G. Geomorphology and Late Quaternary development of Middleton and Elizabeth Reefs. *Coral Reefs* **23**, 249–262 (2004).

46 Choat, J.H., van Herwerden, L., Robbins, W.D., Hobbs, J.P. and Ayling, A.M. A report on the ecological surveys conducted at Middleton and Elizabeth Reefs, February 2006. (Report to the Australian Government Department of Environment and Heritage by James Cook University and Sea Research, Townsville, 2006).

47 Kennedy, D.M. and Woodroffe, C.D. Carbonate sediments of Elizabeth and Middleton Reefs close to the southern limits of reef growth in the southwest Pacific. *Australian Journal of Earth Sciences* **51**, 847–857 (2004).

48 Phillips, B., Hale, J. and Maliel, M. Ecological character of the Elizabeth and Middleton Reefs Marine National Nature Reserve Wetland of International Importance. *Prepared for the Department of the Environment and Heritage. Mainstream Environmental Consulting, Canberra* (2006).

49 de Forges, B.R., Koslow, J.A. and Poore, G.C.B. Diversity and endemism of the benthic seamount fauna in the southwest Pacific. *Nature [Nature]* **405**, 944-947 (2000).

50 Poore, G. *Biogeography and diversity of Australia's marine biota.* Vol. The state of the environment (1996).

51 Slip, D., Hindell, M., and Burton, H. Diving behavior of southern elephant seals from Macquarie Island: an overview. In *Elephant Seals* (pp. 253-270). California: University of California Press (1994).

Index

First published in 2022 by Reed New Holland Publishers
Sydney

Level 1, 178 Fox Valley Road, Wahroonga, NSW 2076, Australia

newhollandpublishers.com

A record of this book is held at the National Library of Australia.

ISBN 978 1 92554 686 6

Managing Director: Fiona Schultz
Publisher and Project Editor: Simon Papps
Designer: Andrew Davies
Production Director: Arlene Gippert

Printed in China

10 9 8 7 6 5 4 3 2 1